THE BARCODE

THE BARCODE

HOW A TEAM CREATED
ONE OF THE WORLD'S MOST
UBIQUITOUS TECHNOLOGIES

PAUL V. McENROE

To my wife Tina—without her creative ideas,
spirited encouragement, and constant support
this book would never have been written.

SILICON
VALLEY
PRESS

Published by Silicon Valley Press
Carmel California
www.siliconvalleypress.net

ISBNs: 979-8-9858428-4-5 (hc); 979-8-9858428-5-2 (pbk); 979-8-9858428-6-9 (ebook)
Cover and book design by Mayfly Design
Library of Congress Catalog Number: 2023901160
First Printing: 2023
Printed in the United States of America

CONTENTS

FOREWORD

A truly successful technology becomes invisible. And then, even if you encounter it dozens of times each day, as you do with barcodes, it becomes forgotten.

Today only on rare occasions, do you notice a traditional bar code. It is typically in a place where you had no idea it existed: on a prescription bottle, on the back of a book, on the door of your car. Most often, it occurs when you buy a retail item, and the check-out clerk twists and turns the item to register a barcode you never knew was there: on a candy bar or a potted plant. And you think to yourself, *huh*, and then go on with your busy day; a day whose contours are, in large part, defined by bar codes.

Today, the only bar code that truly captures our attention, are the oddly compelling 3D versions, those square boxes filled with odd and ancient-looking patterns that seem to hold hidden knowledge and syncretic ritual. And, in fact they do: when we capture that image on our smartphone camera—we are rewarded with volumes of hidden information about a house for sale, or concert we can attend, or volumes of repair books.

Soon, this new generation of bar code technology will fade into the background, as is its destiny.

We regularly celebrate the microprocessor or the Internet or the personal computer as the ultimate success story of the electronics age. Certainly they are worthy candidates. But when it comes to its impact on modern life, its contribution to *both* global economic prosperity and quality of daily life, none of them equal the bar code. Our supermarkets swell with abundance, our manufacturing lines run smoothly and the world's supply chains—via rail, truck, ship and plane—work because they are guided and organized by bar codes, by the tens of billions. If those little strips of black and white bars of various widths were to all disappear, the world's economy would grind to a halt, its inventories molder in warehouses, precious goods never reaching their destination. It is not hyperbole to suggest the bar code is as crucial to the modern world as electricity and telecommunications.

But unlike other technological miracles of the last half-century, who knows the history of the bar code?

The Barcode Book finally tells that history. It is long overdue. Better yet, it's a story as compelling as, say, the history of personal computing. It is a story that contains the high-level dealings of a giant and storied corporation, a tiny and colorful design team, the endless frustrations of creating something new and world-changing, conflicting claims of authorship, failure and success, honor and heartbreak.

No one is better suited to tell this story than Paul McEnroe. In so many ways, the story of the barcode is his story. As a young, hot-shot engineer at IBM, this was his project—and he stayed with it, even as he rose in the ranks of Big Blue to senior management and the barcode project slowly made its way to realization. In these pages, we encounter McEnroe as a flesh-and-blood person, dealing with challenges in his personal life even as he fights to assemble his team, convince his employer to stay with the project, win the fight for the industry standard, sell the technology to potential customers and partners—and finally, convince a skeptical public to accept the presence of a space-age box near the cash register full of laser beams.

An added feature of this book is the omnipresence in the background of one of the most storied enterprises in American business history: IBM Corporation. To the outside world, Big Blue has always seemed monolithic, deadly serious, and relentlessly competitive—and never more so than in the era of this book: a place of white shirts and skinny ties, "Think" posters, and the ultimate safe purchase decision for corporate customers everywhere. What McEnroe provides in the course of this story is one of the most illuminating looks ever at the human side of IBM, the singular personalities, the friendships and feuds, the unmatched commitment to leading the electronics revolution. Reading this book, you can understand at last why IBM became one of the most innovative companies in history—and among the most successful. It's hard to imagine how Paul McEnroe and his team could have achieved what they did anywhere else.

In the end, *The Barcode Book* is celebration of the digital age and the entrepreneurial spirit, whether it occurred in a dusty garage or a lonely laboratory in a big corporate industrial park—and of the men and women who had the vision, the courage and the fortitude to take an 'impossible' idea and make it real. And then to change the world.

—Michael S. Malone, author, journalist and podcast host, is one of the world's best-known writers on business and technology.

CHAPTER 1

SIGNPOSTS OF THE MODERN ECONOMY

Ten billion or more times each day, electronic scanners "read" data encoded in black-and-white codes—barcodes—printed on or affixed to physical goods.

These scans are so common, these barcodes so ubiquitous, that we scarcely notice them. They permeate every corner of our lives. They can be found on Mars rovers and ocean bottoms, in operating rooms and particle accelerators, on weaponry and church vestments, on children's toys and giant earthmoving equipment. You are likely wearing some at this very moment.

Ours is an age of inventions—integrated circuits, personal computers, smartphones, social networks—that sweep the planet seemingly overnight, reaching almost unimaginable numbers of users. The barcode is one such invention that has become one of the most commonly used products and services in human history.

Still, despite its success and ubiquity, or perhaps because of it, the barcode is rarely noticed, much less celebrated.

From Los Angeles to Lusaka to Lima, humans encounter hundreds of barcodes each day. And yet, few of us know where they came from. They were just . . . *there* one day, starting in the early 1970s, at the supermarket or department store or the warehouse where we worked. The price stickers shoppers had always known disappeared, replaced by zebra-striped codes. The clerk at the checkout stand was no longer reading price tags and typing those numbers into a big, noisy cash register. That clerk now stood at a computer terminal at the end of the moving track where we placed our milk and eggs and cereal, scanning each item over a glass window with a curious ruby red light. Suddenly the checkout lines at the grocery store seemed to move much faster.

At the back of the store, in the loading docks and warehouses, workers no longer manually recorded items on inventory lists. Instead they were armed with wands they merely waved over items. And like magic, data was automatically entered into an unseen computer, and inventory automatically replenished.

Perhaps not surprisingly, this dramatic new technology also raised fears and doubts. Where did these barcodes, these wands, come from? How did they work? And what would be their impact? Would this new technology take away jobs? And what about lasers? Would that bright light behind the glass blind checkout workers? Was the government planning to tattoo people with individual barcodes to track and control them? No one knew the long-term economic or personal effects.

Decades later, those reactions seem laughable. Indeed, within a couple years of the barcode's introduction, most people assimilated the technology into their daily lives to the point that they noticed it only on the rare occasions when it *failed* to work: when the checkout clerk had to call a stock clerk to do a price check, an act that once characterized shopping.

Barcodes had another, more subtle impact, one we can see only in retrospect. They made it possible to order, track, and process the sale of goods at an unprecedented pace and accuracy—paving the way for the giant supermarkets and big-box stores. The average supermarket these days has as many as 60,000 different products.[1] Retailers like Amazon offer exponentially more. It is impossible to imagine Amazon, with its vast warehouses, its overnight delivery of everything from books to refrigerators, being able to operate without barcodes on every one of its items, envelopes, and boxes.

The barcode is so pervasive, its use so common, that it's difficult to measure. Ten billion scans a day is merely a rough (and likely conservative) estimate, up from only a few years ago at six billion. Even that staggering growth curve is likely shallow as second-generation, two-dimensional "QR" codes (those black-and-white squares on everything from magazine ads to real estate signs, designed to carry far more information than their original two-dimensional predecessors) are increasingly adopted. As we point our smartphones at our TV screens to download the latest fashion or music, there's little doubt that barcode technology is on the verge of an even more awesome growth spurt.

I had the honor of managing the team at IBM that created and standardized a practical, usable barcode technology. We overcame endless technical and cultural obstacles to turn this dream into a real, workable and, most importantly, *practical*

---

1. "Everything You Ever Wanted to Know About Barcodes," by Gary Arndt. *Everything Everywhere Podcast.* https://everything-everywhere.com/everything-you-ever-wanted-to-know-about-barcodes/.

family of products. In the process, we changed the world. And in the end, that is the true test of any new invention.

LINES IN THE SAND

Every landmark tech business seems to come with its own founding myth: the Traitorous Eight and the semiconductor industry; Apple and the Pirates of Silicon Valley; Atari's first Pong machine in a bar in Sunnyvale, California; eBay's Pez dispensers. Some of these myths are true. Others are partly true. Still others are downright false.

The barcode too has its founding story. It began in 1948 with lines drawn in the sand . . . or, more precisely, with an angry Philadelphia grocery store manager.

The manager was so frustrated by the slow checkout process in his store—not to mention the laborious process of affixing a price label onto every item—that he finally took his problem to a dean at the nearby Drexel Institute of Technology. A Drexel graduate student named Bernard Silver overheard the conversation and told a friend and fellow grad student, Joe Woodland, about it.

Woodland, a budding inventor, took the grocer's request to heart, turning it into an obsession: how can we automate the point-of-sale experience? He became so focused on the problem that he put his studies on hold and moved to Miami to work on it. There, as the story goes, he sat on the beach and drew lines in the sand, pondering the issue.

As Woodland recalled:

I remember I was thinking about dots and dashes when I poked my four fingers into the sand and, for whatever reason—I did not know—I pulled my hand toward me and I had four lines.

I said, "Golly! Now I have four lines and they could be wide lines and narrow lines, instead of dots and dashes. Now I have a better chance of finding the doggone thing." Then, only seconds later, I took my four fingers—they were still in the sand—and I swept them round into a circle.[2]

Why a circle? Because Woodland was not only prescient, but practical. A circle, with the lines concentric like the rings of a tree trunk, would be readable from any angle, and thus not require the checkout operator to line up every item precisely in order to be read.

Woodward and Silver spent the next decade testing the idea. In 1952, they were awarded a patent on their data reader. But there was a problem: if a product code reader was an idea whose time had come, technology had not yet caught up to the actual realization of that idea. A prototype built by the two men, using a 500-watt incandescent lightbulb and an oscilloscope, was an impressive achievement, but utterly impractical. It would have required a full-time technician to read and interpret each scope. And the intense light would have eventually blinded the checkout clerk. The idea, if not abandoned, was set aside until the right component technology came along. And there it sat for two decades.

Then in 1970, a tall, thin supermarket executive named Alan Haberman (like that Philly grocer desperate for improved throughput in his stores) used his formidable networking skills to organize and convene an ad hoc committee of his supermarket industry counterparts to create a competition for a new point-of-sale technology that would become the industry standard. In Haberman's Messianic phrasing:

---

2. Ibid.

*Go back to Genesis and read about the Creation . . . God says,
"I will call the night 'night'; I will call the heavens 'heaven.'"
Naming was important. Then the Tower of Babel came along
and messed everything up. In effect, the Universal Product Code[3]
has put everything back into one language, a kind of Esperanto,
that works for everyone.[4]*

All prospective manufacturers interested in providing this universal product identification technology were invited to participate in the competition. The ad hoc committee chose McKinsey and Company to study each of the proposed solutions and recommend the best. McKinsey further chose Tom Wilson and Larry Russell to be the judges.

Over the next several years, my team at IBM worked very closely with these two men, especially Larry (an outgoing and effervescent guy) in refining our proposal. Everyone liked Larry; he proved to be a terrific promoter and consensus builder. We also found McKinsey to be extremely competent, fair, and easy to work with. All in all, the supermarket industry and its ad hoc committee had set up an outstanding working environment for the companies interested in proposing a solution.

Unlike the grocery industry, the retail industry chose to go with the US National Bureau of Standards rather than using an in-house committee and a private company for help in

---

3. Thomas Wilson Jr. was one of the people I referred from McKinsey and Company to work with us on selecting a code. In *Twenty-Five Years Behind Bars*, edited by Alan Haberman, Wilson says that members of the Grocery Manufacturers of America concluded that there was a need for an "inter-industry product code" to reduce the cost of food distribution. The group formed that ad hoc committee of grocery industry executives to cooperatively pursue a universal grocery product identification code. The ad hoc committee first met on August 25, 1970. In April 1971, the committee concluded that a universal product code (UPC) should be adopted.
4. "Ticker Clocks the Billions of Barcodes Scanned Each Day," by Marcus Wohlson. *Wired Business*, April 12, 2013. https://www.wired.com/2013/04/5-billion-bar-codes-scanned-daily/#:~:text=To%20drive%20home%20the%20bar,is%20more%20than%205%20billion.

selecting a code—a controversial decision. To me and many others, it seemed to give up some control. On the other hand, it pretty much ensured at least national acceptance of whatever standard was selected. As it turned out, we were able to work well with the USNBS, and experienced no problems with their selection process. We were lucky; a lot of other emerging technologies over the past fifty years haven't been able to say the same.

Looking back now from the twenty-first century, I am still amazed by the unlikely combination of events that led to our victory.

CHAPTER 2

AN UNTIMELY PROPOSITION

On a foggy spring morning in 1969, I buzzed up Bayshore Freeway to San Francisco International Airport in my 1961 white Porsche Super 90. I'd allowed plenty of time for the journey: I was meeting my new boss at IBM Corporation. We were to fly together to our company's headquarters in New York.

Soon after arriving at the gate, I saw my boss. Garrett Fitzgibbons was a classic IBM salesman with a reputation as an excellent businessman. He had just been named the systems manager for a new venture that would take IBM applications beyond computer installations and custom programming for clients.

I had been working at IBM since 1960. And now I had been asked to head up engineering development for this new business venture. I had been stunned when Jack Kuehler, a senior IBM development executive who would later become the company's president, came to my office to offer me the job. My good luck had continued with the addition of two top-notch employees to my team: Sarkis Zartarian, a workaholic, was the son of an Armenian general. He was unmatched at putting together presentations (though not at giving them). Mort Powell, the first hire, was a trusted, prag-matic voice. They joined the team as, respectively, marketing manager and, reporting to me, engineering manager.

Sarkis, Mort, and I had been working very long hours to devise a viable proposal for this new business venture. Our task was made more challenging by the fact that Sarkis and Mort were members of IBM's Raleigh Laboratory and lived in North Carolina, while Garrett and I lived in Silicon Valley. I had been working in IBM's Advanced Systems Development Division, which was housed in a think tank known as the Los Gatos Laboratory.

Garrett and I had boarded and taken our seats for the New York flight when there seemed to be a delay in closing the aircraft door. I was surprised to hear my name called out over the aircraft's intercom system. I found a flight attendant, who told me to go back through the ramp to the counter at the gate. There I was handed a telephone.

It was Jack Kuehler. Major changes were taking place at IBM, he told me. In particular, Garrett had resigned from the company to become a senior executive at Memorex, a major competitor. Kuehler instructed me to grab Garrett and pro-ceed to a different gate, where we would board a flight to Ra-leigh. Jack would meet us there on the tarmac with a private plane to fly the three of us to New York. There, Garrett would go through an exit interview, and Jack and I would make the new venture proposal presentation to IBM executives.

On our new flight, Garrett ordered a drink and began to enlighten me about these recent events, of which I had been completely unaware. He explained that he had been in the process of acquiring a new home in Raleigh when he realized that he and his family didn't want to leave the Bay Area. So when he was offered a position as a vice president of development and engineering at Memorex, he took it.

Simple enough. Except that it was Memorex, a fierce competitor, which meant he might be taking company secrets there. Hence the sudden revision of our travel plans. Garrett, now *persona non grata*, was going to be escorted out quickly.

Ironically, shortly after Garrett joined Memorex, his bosses there informed him that they had just stumbled upon the chance to hire the most senior and highly reputed development engineer in all magnetic storage media: Al Shugart.

Shugart had just been forced out of his role as head manager of Storage Products Development Engineering at IBM's huge San Jose Laboratory, reportedly for some sort of scandal. The San Jose lab director had been strong enough to force Al out of his laboratory, but not out of IBM, which offered Al a high-level executive staff position at Group Headquarters back in New York. Why such kid-gloves treatment? Well, Al had a huge following of hundreds of engineers who loved him more than they did the company, and IBM was wary of losing such a well-liked and powerful engineer to the competition. The products developed by Shugart, including the "floppy disk," were widely recognized as the most successful and profitable in IBM's entire product line. Further, Memorex was attempting to compete with us precisely with these types of products. No wonder Memorex reneged on Garrett's original offer and threw all its chips at Shugart.

As a consolation, Memorex offered Garrett the job of vice president of marketing. It seemed the least the company could do after wrecking the man's career. Garrett remained at Memorex for several years, later becoming the general manager of media products. By 1975, he had moved on to become an executive at TRW Inc.

So Garrett was in fact leaving, which was most significant news to me. But the bigger story was that of Al Shugart's departure, which shocked the entire computing world.

ON THE CARPET

In White Plains, following Garrett's exit interview, Jack Kuehler and I waited outside the office of the division president to whom we would make our presentation. I heard a very loud discussion emanating from inside that office. Finally the door was flung open and out walked ... Al Shugart. Al slammed the door shut. He then turned toward me, smiling from ear to ear. "Hi Paul," he said. "Good luck in there."[5]

Now it was our turn. The executive was still red-faced from his shouting match with Al. It was decidedly *not* the optimum environment in which to ask for money, and we were about to ask IBM to invest a fortune into an entirely new industry.

With all the enthusiasm we could muster, Jack and I gave our exciting new proposition for IBM to enter the consumer-

---

5. While he was still at IBM, Al often enjoyed an after-work beer with "the boys." They drank at a bar located just across the street from the company's San Jose campus. After he left IBM, he continued to drink there, socializing with the fantastic members of his former engineering team, and providing them with the opportunity to follow him to his new company. This resulted in the largest exodus of engineering talent that I have ever seen. Several hundred engineers followed Al to Memorex. It also resulted in one of the largest intellectual property lawsuits of all time. Al remained at Memorex until he formed Shugart Associates in 1972. He was eventually fired from the billion-dollar company that bore his name. He then spent several years as a commercial fisherman in the San Francisco Bay and as a tavern owner in Santa Cruz. In 1979, he cofounded Seagate, which soon became the world's largest independent manufacturer of disk drives and related components. Hawaiian shirts were the official uniform for all employees. Seagate, the second billion-dollar company he founded, also eventually fired him. Al spent his last years running his dog for Congress, among other adventures. Al Shugart was one of the greatest characters in Silicon Valley history.

transaction "point-of-sale" business. We asked for development funding of $300,000 for year one, $1 million for year two, and $3 million for the third year.

Somehow the executive managed to swallow his anger and pay attention to what we had to say. Our proposal was well received, and we were granted the requested funding.

Jack and I secretly hoped that our idea would prove successful enough to make a moderately sized and profitable new division for Big Blue. We didn't realize (indeed, it was beyond our wildest imaginings) we had just kicked off a high-tech revolution.

The Barcode Age had begun.

CHAPTER 3

AN UNLIKELY BEGINNING

Anyone who knew my beginnings would have been shocked to see me standing in an executive office of the world's most-powerful company, making a multi-million-dollar presentation about the most arcane technology of the era. Even I was amazed as I stood there. I felt a bit like an imposter who would soon be found out.

My biological parents had both come from the coal-mining town of Salem, West Virginia. Coal mining in Appalachia in the middle of the Depression was certainly a hardscrabble life. Before they met, they had each decided to look for jobs in the Detroit area, where there was at least some manufacturing work. Building upon their common past, they began a relationship. Soon my mother was pregnant. Already faced with crushing poverty, the pair decided to place me at birth in an orphanage in Mount Clemens, Michigan, in 1937.

In 1939, at the age of two, I was picked from a lineup in the orphanage and adopted into a loving home in Dayton, Ohio. My adoptive parents were Joseph Mark (Mac) and Sarah Ellen (Sadie) McEnroe. In their almost-twenty years of married life, they had not been blessed with children of their own.

My new father was a barber, with his own difficult past. Born in August 1891, he too had been orphaned at a very young age. He was taken in, not adopted, by a farming family from Holbrook, Iowa, where he worked hard and received very little schooling. As a teenager, he heeded the words of Horace Greeley to "Go West, young man," and he took off. He made it to southwestern Idaho, where he found a job as a barber. He joined the Army at Sandpoint, Idaho, and endured the agony of the trenches of France for the duration of World War I.

My adopting mother was born in December 1899 in Cincinnati, Ohio. She was third of nine children, the last of whom was adopted. Her family had immigrated to eastern Pennsylvania from Ireland during the potato famine. Her grandmother, Mary Ann O'Brien, came with three brothers from County Wexford just before the onset of the American Civil War. All the brothers fought with the Union in the war, which greatly upset Mary Ann, as she did not consider it their war.

One was shot in the eye and killed and the second died of TB. At one point, my great-grandmother heard that the youngest of the brothers, a teenager, was in an army camp at a little town called Gettysburg, not far from where the family lived. My great-grandmother told my mother of going to the camp, marching into a general's tent, and demanding that her brother be released to her immediately. She always claimed that the boy was, in fact, released and went back home with her. It's hard to imagine a general letting a soldier go home with his sister. But then again, he may have been very much underage. Further, knowing the fate of the other brothers, it is at least a plausible story.

At an early age, Sadie and her family moved to Dayton, Ohio. She told me of watching neighborhood boys Wilbur and

Orville Wright fly their "heavier-than-air" machine around the cow pastures on Huffman Prairie.

My maternal grandfather, Thomas Duffy, was a tool and die maker by trade, so the industrial hub of Dayton offered numerous work opportunities. But for some reason, Thomas wanted to settle in the West. So while the family remained in Dayton, he homesteaded 160 acres in southwestern Idaho near the small town of Melba. From time to time, he traveled back to Dayton; sometimes the children visited him in Idaho, where they would ride horses and live the cowboy life. The oldest child eventually moved to Idaho, married, and lived on the ranch with her father. Sadie also made an extended visit to the ranch, taking a horse over the Snake River on a high railroad trestle with no side railings—with no idea when the train might come.

While visiting the ranch, my mother met my father, who had just returned from the war in France. Sadie and Mac were married in August 1920. Mac worked in a barbershop in Nampa for a few years before moving to Portland, Oregon, where he cut hair in the barbershop in the iconic Multnomah Hotel. In 1926, my parents moved to Coos Bay in Oregon. My mother missed her family so much that just two years later, they moved back to Dayton. The couple had saved up enough to purchase a home for just over $2,000.

Sadie's mother, my grandmother, was a costume seamstress and actress helper at a live theater in Dayton. Even though her family had little money, two of her brothers managed to earn college degrees. One became a doctor, while the other played for the early National Football League, where he was notable for being able to throw and kick both left and right. He later became a lawyer.

Sadie and Mac enlisted the aid of those two brothers to help with my adoption at the orphanage. I was so fortunate to have kind and loving new parents. I have only positive memories of my childhood.

My mother was the boss of our family. She was very bright and outgoing. She was largely self-taught. I doubt she finished high school, but she held her own in local intellectual groups with the ladies of the town who held advanced degrees. Perhaps because of memories of her great-uncles in the Civil War, she drove a big Red Cross canteen truck (it dwarfed her; she stood just 4 feet, 11 inches tall) for forty years, well into her old age.

One story my mother told offered a glimpse of my future: one day just before Christmas, when I was a very small child, she found me on my back in the living-room fireplace, using a measuring stick to gauge the diameter of the chimney flue. It seems that I was worried that Santa would not fit down the chimney.

After my difficult start, I enjoyed a happy childhood. Corpus Christi grade school, where I was a middling student, was just a block away. One of my favorite teachers, Sister Leo Margaret, sent me birthday cards until my eighty-first birthday. I loved almost all sports, especially football. As soon as I could manage riding a bicycle loaded with newspapers, I got a job as a newspaper delivery boy. I even offered my customers a special "business route" to get their papers more quickly—my business sense was already starting to emerge.

After elementary school, I attended Chaminade, the only parochial high school in town. It was a long walk and two bus rides to the far side of downtown Dayton to get there. The school was operated by the Society of Mary. The teachers were mostly religious brothers, who were very well educated and dedicated to teaching. My parents, especially my father, engrained in me that attending school was an amazing opportunity for future success in life.

This attitude wasn't surprising: my father was never able to attend school regularly—even grade school. When he did, he was assigned the tasks of keeping the fire going in the potbelly stove and taking care of other children's ponies at a one-room schoolhouse in Holbrook, Iowa. As you can imagine, he was determined that I get a real education. I knew this and did my best to please him.

At the end of my sophomore year, at the age of sixteen, I was able to get a Social Security card and thus a real job. My newspaper route had gone so well that I was given an unlikely job offer: substituting for the full-time newspaper branch managers on their days off. I managed to get Chaminade to give me a schedule of classes that allowed me to leave school and begin working at 2:00 p.m. each day. I first delivered large bundles of papers to the local grocery stores and other newsstands, and then returned to the branch office to receive and distribute between 3,000 and 5,000 papers to the other newspaper boys.

This was a real job and provided me with enough money to save for college tuition. I was too busy working to join my high school athletic teams, but I did join the math club and was captain of the debate team for my junior and senior years. In my senior year, I joined the drama club. And despite the distractions, I graduated as my class salutatorian (second highest academic student). I had been the first in my family to graduate from high school. My parents were thrilled.

But then just as I graduated, my father was diagnosed with a serious cancer and was unable to return to work.

A DIFFICULT CHOICE

Like so many new high-school grads, I was in a quandary as to what course of study to take in college. I considered business, law, and engineering. I managed to earn a full tuition, four-year scholarship to Notre Dame, but it did not include room and board. Moreover, prospects for a job with sufficient income in the South Bend area were not good. I was heartbroken to turn down the offer.

On the other hand, I realized that if I lived in Dayton, I could stay at home, and I already had a good job. I asked for advice from the engineering managers at the local National Cash Register company. They told me that the University of Dayton was a very good school. Further, because the Wright-Patterson Air Force Base was nearby, it might even be a better option for engineering than Notre Dame; the Air Force supported a great deal of engineering research on campus, and many of its world-class experts taught courses at UD.

Not long after commencing my studies in the fall of 1955, I was offered an opportunity to work on an Air Force-funded research project on the University of Dayton campus. This was a great opportunity, especially since most of my classes were in the same building as the research, which allowed me to work in between classes. So I left my job with the *Dayton Daily News* and began working on the Air Force Electrical Engineering Research Project.[6] Happily, the supervisors were the same electrical engineering professors who taught my classes.

---

6. Wright-Patterson Air Force Base was home to the Wright Air Development Center, which operated research centers on the campus of the University of Dayton. One of their test pilots was Neil Armstrong, who later joined NASA and became the first person to walk on the moon.

MENTORS

I was very impressed with the outstanding professors at the University of Dayton, especially in physics and electrical engineering. One of my physics professors, Prof. Rambowski, was a genius who had been a top scientist in Germany during World War II. He was captured by Russia after the war and clandestinely brought to the United States. Now in the 1950s, he conducted research for the Air Force at Wright-Patterson. For me, it was a rare opportunity to see the thinking process of a world-class scientist. He taught us advanced calculus and vector algebra. He used German notation, which was rare in undergraduate education at that time.

One of my more interesting projects was to study the short-term reliability of the vacuum tubes to be used in missile control systems. This was half a decade before President John F. Kennedy set the goal of landing a man on the moon. I was asked to set up a test operation in which vacuum tubes of the era were subjected to severe vibration while operating at full power.

Vacuum tubes were more sensitive to the problem of vibration in missiles than their semiconductor successors would be. In 1956 and '57, however, semiconductors were still in their infancy. Almost all the vacuum tubes failed the severe test. I noticed that the few that passed the test continued to pass several more of the same tests. I concluded that the best opportunity for a successful launch would be to run the vibration test on new vacuum tubes, select the tubes that survived, and put those "used" tubes into the missiles. I was told the Air Force adopted that strategy. Apparently it worked successfully until semiconductors finally supplanted vacuum tubes forever. I was delighted that an important conclusion came from my first research project.

In my junior year, I was promoted to a position managing other student workers. By the end of my senior year, I was directing about a dozen students. The experience in management and leadership proved immensely valuable once I entered my professional career.

One of my most significant experiences at the University of Dayton was as part of the debate team. I served as its captain for three years as we debated many of the top teams (Harvard, Notre Dame, and William and Mary among them) at tournaments around the country. The experience of public speaking while thinking on my feet was great preparation for presenting to industry forums, corporate executives, and at other events. Interestingly, in all my many debates, I never met another debater who was an engineering student. They were mostly pre-law, business, and English majors. At graduation, I was proud to receive a special award from the University president for Excellence in Debate.

I graduated in 1959 as the valedictorian of my class. The honor was nice, but my dad's attendance at my graduation ceremony was one of the most moving moments of my life. Because of his own lack of education, he had engrained in me the importance of a college education. He was in poor health and had to be escorted into the large auditorium in a wheelchair. Because I was giving the valedictory address, he was seated between the president of the university and the guest of honor and speaker: Robert S. Oelman, president and CEO of National Cash Register Company. My father was immensely proud of me. From that day on, he felt his life was complete.

Even before graduation, I had researched what graduate school I wanted to attend, and what course of study I wanted to take. The subjects of greatest interest to me were still engineering, business management, and law. The City of Dayton Engineers Club presented me with an interesting opportunity in law, offering me a full scholarship to Harvard. Their interest was getting young people to study patent law. While I thought business law could be very interesting, I had little interest in a solitary career reading and writing detailed patents.

The Engineers Club generously told me that I could switch from patent law to another form of law after one year. But it didn't seem fair to betray their desires. In addition, much of the advice I received from interviews with successful managers and professors recommended a graduate degree in engineering as the best entry to a technical business management career, especially if I later added an MBA to my qualifications.

Finally, I decided to build on the fact that I'd just completed four years of school in electrical engineering and three years working in the Air Force/university research laboratories. I sought out an electrical engineering graduate school. Ultimately, I decided to apply to two programs that were fairly close to home, and two on the West Coast: Purdue, Illinois, Stanford, and Caltech.

I quickly heard from all these universities except Purdue. Each offered me a fellowship in electrical engineering that completely covered tuition, miscellaneous expenses, and a generous living stipend. I was torn between Stanford and Caltech. Finally I chose Stanford.

My parents and I were extremely excited by my decision. But as I prepared for this marvelous opportunity, my father took very ill. Doctors predicted that he would not likely survive the next couple of years. Stanford's offer was so generous that I could have attended school there without any financial worries, but traveling back and forth to the Midwest was not an option.

Then, just after receiving the bad news about my father, I finally heard from Purdue. Its offer was similar to that of the other universities. Specifically, the president offered to make me a Purdue University Fellow, which meant full tuition, miscellaneous expenses, and room and board in the new graduate student living quarters—all with no research or teaching responsibilities.

From Purdue, I would be just three hours by car from home. From Stanford, it would be a long and unaffordable plane trip, as jet airliners had not yet made the scene. I had already accepted Stanford's wonderful offer. I decided to seek the advice of that school's dean of engineering, Dr. Joseph M. Pettit. Dr. Pettit took the time to talk to me in order to understand my situation, and he offered me solid advice. He gave me a new offer that would completely solve my problem. He told me that personal and family relationships were of utmost importance, and that Purdue was a great school. He recommended that I accept its offer, and he promised that he would make sure that Stanford would renew its offer to me for its PhD program.

I was deeply touched by Dr. Pettit's understanding and by his offer. So I accepted Purdue's Fellowship and planned for a future at Stanford.[7]

---

7. In 1972, Dr. Pettit became the president of Georgia Tech and has since been recognized for leading that institution to a position of national acclaim. No surprise to me, as his dealing with me demonstrated his care and compassion. He provided a magnanimous solution to a person who was not yet even a student in his university. I will never forget his action on my behalf.

BOEING

During my senior year at Dayton, I had applied for a summer engineering program at Boeing Airplane Company. Upon acceptance, I drove to Seattle. There I was assigned to the Bomarc missile program, a supersonic ramjet-powered, long-range surface-to-air missile (SAM). At the time, it was the only SAM ever deployed by the United States Air Force. It helped that I had already been working under contract for the Air Force for the best part of my four years at the University of Dayton.

Working at Boeing was a memorable experience. From the immense gymnasium-like office that housed hundreds of engineers, I could see out the window to the Boeing airfield where the newly developed 707, the world's first commercial jet, was being tested. As I watched, some 707s took off on very steep paths while others seemingly just barely made it off the ground. It made for very interesting window-side coffee breaks for me and my fellow Bomarc engineers.

PURDUE

In the fall of 1959, I left Boeing to begin graduate school. At Purdue, I found excellent programs, top-notch professors, and impressive academic rigor.

In my Advanced Electromagnetics course, the professor began by saying he thought it was important to go back to the beginnings of electromagnetic theory and study directly from Max Abraham, a student of James Clerk Maxwell, who was the father of the field.

There were only six students in the class. All of us descended upon the university bookstore to procure the text,

Classical Electricity and Magnetism (1932) by Abraham and Becker. We were shocked to find that it was available only in German. Even the equations were in German notation. That made it particularly difficult for the *other* five students. Fortunately however, I had become familiar with that notation in the class taught by Prof. Rambowski at Dayton.

By the spring of 1960, I began to home in on a strategy for completing my master's degree, getting a job, and, if possible, working on a PhD. I have to admit though, I was getting tired of school and increasingly interested in marriage and raising a family.

I began the ritual of corporate interviews. My focus was on four companies, all in the western United States, a region I'd come to love: Boeing Airplane Company in Seattle; Los Alamos National Laboratory in Santa Fe, New Mexico; Honeywell in Colorado; and IBM in San Jose, California. The IBM interviews were especially interesting in that the interviewers were engineering managers with job opportunities in their home organizations. I had to turn down multiple invitations to visit the company's many different sites in New York State before I finally received a call from an engineering manager in San Jose.

In the end I took that job. It seemed to fit me perfectly, in that it included a program of study at nearby Stanford University—where, as noted, I already had a grad school spot waiting.

GO WEST

A few days after graduating from Purdue, I married Ann Margaret Rawers. We had dated in high school and college, and she had just graduated (in three years) from the University of

Dayton in education. A few days after our wedding, with $600 to our name, we drove west for new adventures in California.

We quickly found a new apartment in Campbell for $79 per month. Two days later, I walked through the doors of IBM San Jose.

It had been an unlikely journey getting there, and what was to come felt just as unlikely.

CHAPTER 4

BIG BLUE

As we prepared our 1969 presentation for the IBM executives, Jack Kuehler and I considered many potential market opportunities: banking, manufacturing, insurance, airline reservations, and distribution. With its large mainframe computers, display systems, and the requisite communications to interconnect these various systems, IBM commanded a dominant position in *all* these industries. After all, in those days, IBM was the most powerful technology company in the world. In those early stages, we were essentially agnostic—prepared to go wherever Big Blue wanted.

On the other hand, IBM already had specific application products that addressed the most lucrative opportunities in many of these businesses. Banking and distribution, however, were exceptions. Or so we thought. It turned out that a significant stealth effort to develop new products for banking was already underway at our Kingston, New York, laboratory.

We learned that IBM's most recent distribution development had failed to muster the cost and function requirements to proceed. Further, we could see from our own studies that the distribution industry—in supermarkets, mass merchandising, and general retailing—had identified a special need for item identification. In particular, the industry needed faster customer service at checkout and to lower operations costs through automatic inventory reordering, streamlined warehouse management, and the elimination of price marking.

How were we able to understand the requirements of all these different industries in the short time before our presentation? And how were we able to detail the specific needs in the overall distribution industry?

The answer is that we were blessed with IBM's incredibly efficient sales team, which in those days was in constant conversation with customers about their current and future needs. At the time, IBM installations satisfied more than 90 percent of the customer's data-processing requirements in these industries around the globe. IBM's Data Processing Division's (DPD) sales organization had documented these conversations quite thoroughly, and we had access to all of them.

The documentation of opportunities in supermarket, mass merchandising, and retail was particularly well defined. It helped that the item identification, point-of-sale, and inventory-control requirements of these customers were well detailed in the various trade magazines of the time.

Our trio had an almost perfect background for preparing this particular proposal as it turned out. Sarkis, armed with a bachelor's degree in engineering from MIT and an MBA from Harvard, was brilliant, hardworking, and relentless. His ability to analyze business opportunities was outstanding. Mort was an experienced engineer and engineering manager who had developed numerous IBM products and overseen their manufacture. I had been working in advanced product development for the previous nine years.

I will briefly describe the general nature of the group I

worked with to provide a glimpse of what life was like at one of the most innovative and successful companies of all time.

It had been nine years since I joined IBM's Advanced Systems Development Division in San Jose. Put simply, the division's goal was to identify significant computer application needs that could not be addressed by existing equipment. That charter required us to work on complex systems to develop the technologies needed to supply those missing components.

As a result, I was blessed to be part of many exciting and groundbreaking research projects. In particular, I spent the first couple of years developing scanners. I had a wonderful manager (Bob Schneider) and lead engineer (Carroll Brown) to guide my initial work.

On my first day at IBM, I was assigned a desk in a two-person office, one side of which was all glass with a beautiful outdoor patio well bordered with vibrant green plants. But the best aspect of the office was the other engineer with whom I shared it. His name was Jack Kuehler. He became a lifelong friend. Jack mentored me through the politics of Big Blue. He knew his stuff: he was destined to become president of IBM, by then the most valuable company in the world.

Jack was a mechanical engineer. Although we shared an office, we were not assigned to work on the same project. Instead, working under Schneider and Brown, I was tasked with developing a high-resolution flying spot scanner for reading micro images. The images were optical, encoded on film rather than paper. The application for the device I was to design was the automatic reading of large engineering documents; maps, especially oilfield maps; and Central Intelligence Agency (CIA) images stored on the "head of a pin"—microdots. The last were of the type Ian Fleming's "Q" could have offered to James Bond. The light source for the reader was to be a Litton high-resolution cathode-ray tube (CRT). Happily for me, this project proved to be the best possible preparation for facing the problems of scanning optical codes, such as a barcode.

After the scanning project, I moved on to character displays, using special analog circuit design. Again I was lucky: this technology happened to be the focus of my current coursework at Stanford under Prof. Malcolm Myers McWhorter, and buttressed by semiconductor studies under Prof. John Moll, famous for his Ebers-Moll equations and semiconductor models.

In 1964 the Advanced Systems Development Division built a new laboratory on Guadalupe Mines Road in Los Gatos, California. Only a few miles from its original location in San Jose, the new lab was one of the finest and most singular development facilities anywhere in the world. Rising on a gentle slope within the 100-acre site, the building was nestled just below the live oaks that blanketed the steeper portion of the property.

The entire laboratory was built of redwood. The offices were designed for one to three engineers and located in a multitude of Maltese crosses that formed the perimeter of the building. From within these office windows, employees overlooked the rolling hills to a golf course below.

The laboratory had been the creation of Reynold B. Johnson, the lab director. He was already well known for his instrumental role in the development of RAMAC (the first disk drive) and other notable IBM products. These products played a key role in IBM's unmatched success during this era. The lab itself enjoyed an outstanding reputation and was fondly known as the Los Gatos Think Tank. (Sadly, on a trip to Los Gatos some years ago, I drove past the site only to find the fabulous laboratory gone, replaced by residential homes.)

Upon moving to the Los Gatos laboratory, I began using new fiber-optic technology to develop fax machines. Having proven the feasibility of a mechanical fax machine, my target application switched to optical recognition of codes and characters on paper, as well as printing on paper with a laser light source. The most exciting of these applications was the scanning and decoding of the ZIP code on envelopes for the United States Post Office. Unfortunately, the system worked well, but the post office ran out of money. (Fifty-five years later, their inability to have implemented a state-of-the-art technical solution does not seem surprising.)

I began work on magnetic projects. One included putting a magnetic stripe on a small plastic card for a joint experiment with Lloyd's of London in England. That particular early 1960's project ended in failure when Lloyd's concluded it unfeasible to expect enough people to carry a plastic card in lieu of cash. I guess you can call that "missing the boat."

Another interesting project included putting a magnetic stripe on the tickets issued at airport ticket kiosks. In theory, airline customers could purchase and print tickets themselves for shuttle flights between major airport hubs, such as Chicago, Boston, New York, and Washington, DC. This project ultimately met with some success: our experimental equipment was installed in each of these airports and used by the public to purchase tickets between these cities.

We also undertook a magnetic project to provide a ticketing system to the new San Francisco Bay Area Rapid Transit Authority (BART). We built experimental equipment that printed magnetic tickets, as well as a reader that could be placed at the station gate. In practice, a BART rider could place the newly printed ticket into the reader for entry. As far as I know, the airline kiosk tickets and the BART tickets were the first applications designed for consumers that used a magnetic stripe placed on paper stock. That BART ticketing system, which we installed in the 1960s, remained in use for decades, and was still working at the dawn of the Internet Age.

As you can see by the nature of these products, our focus was not on basic research, nor even in potential applications, but on practical technology to meet existing, real-life needs. That proved crucial in our creation of barcodes and their readers.

Later, another IBM San Jose group undertook the huge challenge of developing the biggest digital file the world had ever known. Its primary application was for atomic physics laboratories. A photo-digital type of file was chosen, and I was asked to help in the scanning part of that task. Several terabyte photo digital files were built and used for a decade by the National Atomic laboratories. (I describe this significant system in more detail later.)

My final task before the barcode project was to develop a computer-aided educational classroom. The idea was to provide very inexpensive keyboards and displays at each student's desk—much like putting a personal computer on each child's desk. In this case however, the "PC" was a simple terminal driven by a backroom computer controlled by the teacher. This was twenty years before the true personal computer entered the scene.

A PERFECT STORM

In retrospect, the barcode reader almost seems predestined. All the technologies, especially scanning, that I worked on leading up to it contributed in some crucial way to the development of a state-of-the-art, low-cost, point-of-sale system with item identification.

I grew up in a world where cutting-edge, point-of-sale technology took the form of big, beautiful, heavy, cast-iron NCR cash registers, with their scores of levered keys, cash drawers, and up to six printers. Black-and-white numbers popped up behind a glass window as a cashier pounded on those keys. These registers were indeed a beautiful sight. All of a sudden though, I saw them in an entirely new light. To me, they were akin to a large, ancient ship in the middle of the ocean. And all the technologies I had worked on over the previous ten years swirled into a Perfect Storm. That storm would send the ship to the bottom of the sea, replacing it with optical and magnetic codes, handheld scanners, checkstand laser scanners, integrated circuits, LEDs, ticket printers, contact recording disks, and fail-safe systems for management and inventory control. That ancient ship did its job for almost a century, and it alone had controlled all the seas in that century. Now its POS replacement inherited all the oceans of opportunity.

It would take years of work, but my experience told me it was all doable. From this point forward, I knew we were on the brink of an entirely new age of general retailing throughout the world. I was intimidated, but I was a loyal IBM developer. Whatever Big Blue needed, I would do it, and I would do my best.

ENGINEER

So there I was, with a pretty broad electrical engineering education and ten years of development experience, being invited to start a new business venture in IBM. What a dream, especially having just completed my "engineer's MBA" at Stanford's graduate school of business, where similar opportunities were likely years in the future for most of my fellow students.

During my time at IBM, I attended Stanford part-time. For the first few years, I focused exclusively on courses in the graduate school of electrical engineering. These, coupled with my graduate work at Purdue, completed the coursework required by Stanford for a PhD in electrical engineering.

To complete the PhD degree, Stanford required at least one year, full time, at the university. IBM was happy to pay my tuition and allow a flexible schedule, so I could take courses during the day and make up worktime in the evenings and on weekends. The full-time requirement, however, would necessitate my taking a leave of absence from the company, which would have caused two serious problems for me. First, I was a young man starting a family and would have no income for that year. Second, I would have to give up my exciting IBM work assignments in the hope that I would be lucky enough to find equivalent new assignments after that lost year. I had seen others take leave for full-time education, only to return to IBM to find the best jobs filled by those who had remained.

When I discussed this problem with my Stanford advisor, he advised me to take advantage of a new Stanford program that offered a course of study in both business and engineering, leading to the degree of Engineer rather than a PhD. (That advisor, Prof. John Linville, is generally regarded as one of the founding fathers, after Frederick Emmons Terman, of Silicon Valley.)

Prof. Linville pointed out that under this new degree program, I would be able to enroll in almost any course in the

Graduate School of Business. And better yet, I had already completed more than the necessary engineering coursework required for this new degree. Some even thought that a PhD could be a disadvantage in pursuing a career in senior IBM management.

Mine was a surprisingly easy decision: I decided to pursue the new program in business, focusing on marketing and management studies. Serendipitously, a couple years later, in 1969, just when I had completed all that business coursework, IBM began searching for someone to lead the development of its new startup venture. I cannot say for sure, but I believe my engineering experience, coupled with my more recent work at Stanford in business administration, led IBM to select me for the role. In any case, it didn't hurt. I thanked the company for its trust in me by using that venture to embark on the creation of the barcode.

CHAPTER 5

ASSEMBLING THE TEAM

In spring 1969, walking with Jack Kuehler out of our successful point-of-sale presentation, I was briefly exultant: *we'd done it!*

But that emotion didn't last long. Feeling a bit like Caesar crossing the Rubicon, I thought, *Well, we won the battle. Now it's time to put detailed business and development plans in place.* IBM's management may have been friendlier than the Roman Senate, but to me, the scenario was no less risky. At Big Blue, then and now, most developments are improvements of existing product lines. Our initiative, however, was a complete start-from-scratch.

The first question Jack and I faced was: where do we do the development? At the time, IBM had nearly two dozen development laboratories around the world, more than half in the United States. It turned out that one of the company's newest development labs was located in the Research Triangle Park of North Carolina. (The Park is located within an isosceles triangle whose baseline runs from Chapel Hill to Durham, with Raleigh as the apex. Each of these three cities has a major research university: University of North Carolina, Duke, and North Carolina State University, respectively.)

North Carolina promoted the area to companies looking to take advantage of research opportunities provided by these outstanding universities. It offered significant tax benefits to companies to relocate there. Around 1965, IBM acquired 600 acres and built a significant development laboratory and manufacturing facility in the Park. At the time, not enough IBM projects were being developed at that lab to fill the manufacturing line nearby.

In light of the gap, we decided to move our systems and development headquarters to the Raleigh laboratory. The name given to our group was Consumer Transaction Systems—or, as it came to be commonly known within IBM, CTS.

CHOOSING THE MARKETS

In our quest for funding from IBM executives, we asked to be responsible for all development for any products that would aid consumers in transacting with any type of retailer. We included in that request supermarkets and mass merchandisers, major retailers, and dry-goods stores—including the big three: Sears, Penney's, and "Wards" (Montgomery Wards)—fast-food stores, gas stations, and hotels and motels, among others. Banking was a notable exception.

This was an ambitious request. Further, at that moment, we clearly didn't know enough to back it up. We still needed to do detailed studies of any competitive products currently in use by the consumer and retailer. The primary supplier in this marketplace, since 1884, was NCR: the National Cash Register Company.

As it turns out, in another bit of serendipity, I was more familiar with NCR than anyone else in the group. I had grown up in the shadow of the company's headquarters in

Dayton, Ohio. The fathers of many of my childhood friends worked at NCR. As a result I was a frequent guest at the company's extensive recreational facilities, including a fabulous park known as Old River. I even knew the president, and later CEO, of the company: Robert S. Oelman. He had given the 1959 commencement address at my University of Dayton graduation ceremony, and we were seated together during graduation week at several events and meals because I was the valedictorian of the graduating class that year.

Like many Daytonians of the time, I was in awe of NCR. Its many thousands of employees and acres of buildings were like a citadel of engineering, filled with an army of engineers. NCR controlled more than 90 percent of the cash-register marketplace, now being called *point-of-sale*. But in 1969, I saw in that strength an opening into which we could slide an entirely new type of point-of-sale terminal, dealing a fatal blow to old-fashioned cash registers.

That opening was in the point-of-sale device itself, and the fact that it did not fit into a complete system. NCR's existing product consisted of a cash drawer controlled by a keyboard with a mechanical display and as many as six printers. Very sturdy, very reliable, and it weighed a ton. Most important for us, it could not be easily adapted for use with any existing technologies of item identification.

The result was that NCR's solution, the status quo for the industry, was a technologically obsolete device in use at almost every point-of-sale location throughout the world. Undaunted by this challenge, I considered this an almost unheard-of opportunity. NCR was a dinosaur; it had grown complacent in its market leadership, ignoring the arrival of the information age. The dinosaur was ripe for replacement. Meanwhile, other potentially competitive devices of the time were either low-volume or still experimental, and thus did not present a serious threat.

BUILDING ON FAILURE

As I discovered, IBM had previously taken a couple of passes at the point-of-sale marketplace. None of these initiatives resulted in a product offering, although a custom systems experiment the company conducted in San Jose shed some light on using a POS device in an item-identification application.

The most recent significant IBM point-of-sale development had taken place in the Rochester, Minnesota, laboratory a few years before we got our green light from management. The good news for us was that IBM's business model required realistic projections of volumes and profit, which the Rochester program was unable to provide. The bad news was that our new development had to meet the same requirements.

In the late 1960s, the Supermarket Institute and the National Retail Merchants Association had both been calling for item identification, believing it would speed up checkout and reduce costs in warehousing and restocking. The reports generated by these two industry associations proved to be a wealth of information for our team. Less exciting was the fact that these new requirements were well documented in the industry-oriented magazines of the day—meaning that entrepreneurs everywhere might be heading down our same path.

PRECEDENT

If the big old cash register was not easily adaptable to the current industry requirements, what technology was required instead? Perhaps more importantly, what would be at the heart of this new item-identification technology? Finally, could that new technology be used in supermarket *and* retail applications, thus making it financially feasible?

In answering these questions, I called upon the varied experiences of my first nine years at IBM. This proved to be a great strength. Of particular value was my assignment to the Advanced Systems Development Laboratory, where I evaluated different technological product applications, determined what technologies were missing or deficient, and finally tried to perfect and implement those technologies. As a result, I found myself working in new and different technologies every couple of years. Better yet, most of these technologies were related to the human-machine interface that was so relevant to the new point-of-sale requirements.

This history was why, in our proposal for this new startup, we asked for and received approval to proceed on the basis that we would be treated differently from IBM's other standard development programs. We knew we could not come out of the starting gate meeting IBM's standards for profitability. Fortunately, IBM senior managers also recognized that was the case, so they let us paint an imaginary red line around our operation, enabling us to operate differently. We would have a lot more freedom in outsourcing components from other manufacturers, and also have our own dedicated sales force. Finally, as noted, we did not have to achieve a very low ratio of development expense-to-product revenue.

Given our new, and singular, ground rules, I thought it important to constantly advertise—both inside the company and to the outside world—that we were in fact doing business as a Silicon Valley-type of startup rather than a classic IBM development.

As I inquired about space available to house our develop-

Paul McEnroe at his desk in 1974 in the "not quite Skunk Works" Raleigh building where the Barcode was developed. (His kids did not grow up to be artists.)

ment, I became quite excited about a reasonably large, older brick building in downtown Raleigh that was being vacated by another development organization. This building was about twenty miles from the main IBM laboratories in the Research Triangle Park, also was in an industrial area of Raleigh, adjacent to freeway ramps and convenient to Raleigh's residential areas, where most of our future engineers were expected to live. While it wasn't exactly like a Lockheed "Skunk Works" or a Silicon Valley "garage," this site at least conveyed the message to visitors that we were running a different type of operation. More importantly, it induced a feeling of thriftiness and independence in the team it housed.

I rented the space.

HOME FIRES

Meanwhile, in San Jose, Ann and I now had two children, Maureen and Paul Jr., and we were expecting another. Ann was teaching at a high school. We had a nice tract home in the still-rural Almaden area. IBM had an excellent employee relocation program, which covered all our moving costs. Ann did not like to fly. Knowing we had very similar tastes, she convinced me that I should just pick out a new home on one of my business trips to North Carolina.

In the end, I chose a newly constructed and almost finished Southern Colonial in Raleigh, located on a large wooded lot adjacent to community parkland of quite a few acres, the hub of which was a wonderful swimming pool complex. I was too busy with my startup to fly home and join the family for the move. So Ann and the kids took the California Zephyr train from San Francisco to Chicago and then on to meet me in Raleigh. Fortunately, Ann loved the house, and our kids loved the pool.

TALENT SEARCH

During this period, my most critical job was to hire a core staff of engineers.

The first engineer I hired was Mort Powell. He had already come out to help with our initial planning in San Jose. Not long before, Mort transferred from IBM's Endicott, New York, Laboratory to the Raleigh Laboratory. This mattered because several other excellent engineers who spent most of their careers in Endicott also moved to Raleigh. Jerry Harries, a good friend of Mort's, was director of the Raleigh lab, and that connection proved valuable. Mort became my first engineering manager.

The next additions to the team, in alphabetic order, were Herm Baumeister, Bill Betts, George Laurer, and Alex Sawtschenko. Bill and Alex (the former almost old enough to be the father of the latter) were very experienced communications systems designers who managed to work well together. Laurer, a neighbor of mine and a classic nerd, was a brilliant senior engineer with a great deal of experience in many fields, including communications, mechanisms, and printers. Baumeister was a talented general-purpose engineer. He and Laurer were very senior engineers, much older than me, and were hard to manage because they had their own way of doing things and didn't want to be led by the new kid.

In fact, one reason why so many of these talented engineers were available was that they had been considered "difficult" to manage—a description that actually meant they were individual contributors, rather than managers. With their years of experience and competency, they were well paid, and their salaries were well earned. But they often earned more than most of their fellow engineers as well as many of the younger managers to whom they reported, which could be a difficult situation for those young managers. Thankfully, since I had been given a significant multi-level promotion to take this assignment, and since Mort was already a high-level manager with equal or more experience than all the others, neither he nor I had to deal with this salary/experience mismatch.

Unfortunately (perhaps fortunately in the long term), none of these veteran engineers had any specific experience with point-of-sale devices, nor were they particularly familiar with the latest details in circuit design, especially with integrated circuits (which were just coming into their own). That meant we had a slow start as these engineers taught themselves this new subject. But it also meant that they were not constrained by prior art—no one had told them what *couldn't* be done. That proved to be a crucial advantage.

In reviewing the post-mortem of the failed Rochester Laboratory cash-register project, I came across a brilliant young engineer named Roger Kaus. In the final stages of that earlier project, Roger proposed a more modern solution. I became convinced that he might become a key designer of our point-of-sale terminal. He was reluctant (perhaps because of the failure of that project), but I pulled out all the stops to convince him to move to Raleigh and join our team. He did, and his contributions proved outstanding.

Now we had our team. This founding core of seven engineers (including myself) proved to be almost all the engineers we needed for our first year.

In any startup, keeping the funding people happy is a requirement. In this case, the funding people were not venture capitalists, well-heeled angels, or partnerships, but rather IBM executives. But just like those other types of investors (perhaps even more so) if they didn't like what they saw, they could simply pull the plug.

For our team, the funding people were: the lab director, the division president, and the group executives. I was the systems development manager, and since Garrett Fitzgibbons left the company, I was also the acting systems manager. In that position, I reported to the lab director, Jerry Harries. Jerry reported (my good luck) to the division president: Jack Kuehler. Reporting to me in the systems manager role were: the marketing manager, Sarkis Zartarian; and the business manager, Bob Dean. My job was to keep all the individuals above me happy with our progress, and those below me happy with our direction.

The best way to please both constituencies was to come up with a product-offering plan that was economically justifiable and that functionally addressed the customers' needs. It was equally important to keep senior management convinced of the viability of the development program.

Sarkis, Bob, and I focused on the latter issue daily. We flew frequently to IBM's executive offices in Harrison, New York. IBM managers were constantly called on to present their opinions and recommendations to senior executives. In those days, the standard medium for those presentations was the flip chart. For younger readers, flip charts were proto-PowerPoints: giant pads of blank paper hung on an easel. Using colored markers, the presenter drew diagrams, bullet points, or lists. Even IBM CEO Frank Cary, to make a point to a board member or another executive, quickly scribbled it out on a flip chart, ripped the chart off the easel, and scrambled down the hall of "executive row" with the partially crumpled chart under his arm to make his case. Goodness knows how many flip charts we used to make our case to IBM executives.

These presentations took a lot of precious time from our fledgling startup. Nevertheless, in preparing those charts, our team gathered huge amounts of information about what we needed to be successful in our newly defined Consumer Transaction Systems market. Based on that information, Sarkis talked several regular IBM salespeople, who until then had specialized in selling computers to major supermarket and retail chains, into joining our market-planning team.

Their contacts and experience proved extremely valuable as we defined the parameters of our design.

While we were quite successful in setting up our business plan and getting senior management to agree with our direction, we were still missing a permanent systems manager. I had been acting in that role, but my duties as the systems development manager required a full-time commitment. For that reason, I was extremely happy when Jack Kuehler announced that our new, full-time, systems manager was J. Roger Moody, a friendly, effervescent, experienced, gung-ho salesman and manager. Roger joined the team immediately.

"Sales" and "marketing" had distinct meanings at IBM. "Sales" meant the selling of the product to a customer. "Sales management" meant the organization and direction of the team of people who actually sold the product. "Marketing" meant defining the characteristics the product required to meet the needs of the customer. This included coming up with a price. Marketing had to work closely with engineering, which took the product requirements from marketing and developed the best possible product that current technology would allow. Marketing also worked closely with sales to ensure that the latter could sell the product at the proposed price with the functionality and performance that engineering was able to achieve.

In our CTS organization, the systems manager (my acting role) and his team were within a business development operation: the Systems Communications Division (SCD). Sales management, on the other hand, was in the Data Processing Division (DPD). The DPD was responsible for selling all IBM products, except for typewriters and small office systems, which were referred to as "low end products."

For "high end products," IBM had several other devel-

opment divisions besides SCD, each of which had development as well as profit and loss (P&L) responsibility for their
different product lines. These divisions, as well as the sales
division, all reported to the group executive of the Data Processing Product Group, or DPPG. The DPPG group executive
reported directly to the management committee, which included the president and CEO of IBM.

Now that our request for a dedicated sales force was
granted, we needed a manager. That manager would report
to John Akers, the DPD executive who headed up the distribution industry organization. An ex-fighter pilot, John was a
good friend of CTS in that he had been very supportive of our
startup proposition. He later became IBM president and CEO.
John chose Marvin Mann, a very experienced, high-level IBM
sales executive. Marvin (in New York) hired a marketing manager, Bill Carey, to work with him from our Raleigh office.

The result of all this was that our sales and development
organizations reported up through two different divisions
to a group executive. This might have led to confusion and
conflicting messages from the top, but it worked out fine. My
earlier concern about having a separate sales force was based
on wanting to be sure that our individual salespeople could
sell only CTS products. I believed that if they could also
sell other IBM products, they would make so much "easy"
money selling computers that they would not be motivated
to sell our point-of-sale systems. After all, the latter were going into a new application and thus would be a much more
time-consuming and difficult sale. And in fact, this concern
proved to be true. We probably would not have been nearly
as successful had the salespeople who sold our products also
been able to sell computers.

LETTER TO COUNT

From our development team's work with the sales and planning groups, it quickly became apparent that supermarkets and retailers required different systems solutions, hardware *and* software. That said, in terms of the underlying technology, we concluded that item identification was key to both applications—with some wrinkles. For supermarkets, item identification required ten numeric digits, plus an additional digit or two for country codes. In retailing however, thirty-six alphanumeric characters were required. The extra numbers and characters would encode all the detailed information that merchants needed. For example, in the case of fashion merchandise, such as dresses, the code might include the color, size, material, designer, manufacturer, and price. Further, the speed of reading each item to be identified was much more important in the supermarket world, simply because of the larger number of items in an average purchase.

For supermarkets, the printed price would have to be unreadable by customers to allow for rapid price changes. However, without a price on each item, if the system ever failed, the checkout clerks would have no way to determine a price. Thus, we added a supermarket system requirement to provide fully duplexed controllers, so that a failing store controller was automatically backed up by its duplexed counterpart.

The above feature was not necessary in most major retailing applications where prices don't change as rapidly. That's why, other than the core reader technology, we created separate system designs for the supermarket and retail industries. That said, many system components (including the terminals themselves, the communications system connecting the terminals to each other and to a store controller, the store manager's terminal, and the warehousing and headquarters host computers) remained quite similar for both customer types. The large number of shared components and system designs significantly reduced the overall engineering, product, and servicing costs.

TWO PATHS

We very quickly reached a fork in the road regarding our two target markets. Supermarket required only numerics and not alphanumerics. As well, the cost of applying the code to products needed to be very low. These factors led us rather quickly to choose an optical code for this market. Retail, on the other hand, with its alphanumerics and more flexibility in terms of the cost of applying the code to the product, pointed us in a different direction: namely, to provide a magnetic code.

This difference in the item identification technology required by the two different applications, along with supermarket's "fail safe" need for duplexed systems, drove me to set up separate engineering groups. The plan was still to share many systems components, but have independent engineering managers responsible for success in their respective applications.

We confirmed that we could, as planned, provide a single systems design, with a common communications system that tied together all the various system components noted above. This proved to be one of the world's first multilevel systems with programmable intelligence in each of the many levels of its overall system design. I asked David Mackie to lead the supermarket, and Len Felton the retail systems programs, respectively.

With the organization set, we were now ready to get down to the nitty-gritty of detailed engineering design.

FINAL JUDGMENT

While the systems design was critically important to the basic success of the entire program, it was also necessary to have a reliably readable code for retail and supermarkets that would pass the test of time.

Automatic item identification was the primary function (the "killer app" in modern parlance) that would justify this rather expensive and complex three-level system. That's why the Supermarket Ad Hoc Committee hired McKinsey to be the interface to the technical manufacturers. McKinsey was, after all, one of the leading consulting companies in the world, and was hired because the system was too complex for the supermarket executives on the Ad Hoc Committee to understand its operation, much less decide which of several solutions was the best.

The committee's specifications (which I discuss in depth later) included requirements for a numeric-only, optically readable code that the scanning equipment automatically recognized. Further, the scanning needed to occur as an item was being pulled across the checkout stand at a velocity of up to 100 inches per second. Additionally, that optical coding must not take up very much space on the package. Two factors drove the minimal space requirement: first, some packages are very small; second, grocery manufacturers regarded all the space on even larger packages as critical to their advertising.

The Supermarket Ad Hoc Committee's competition, dreamed up by Alan Haberman, was announced in 1971 and continued until August 1973. More than a dozen companies showed an interest in code development, or manufacturing systems to read that code. We looked into many different types of codes discussed and proposed. I have no doubt our competitors did the same thing.

One type of code we found interesting was a "bull's eye," being tested and proposed by RCA. A variant of this design was also being proposed by Zellweger of Switzerland. This code used various thicknesses of circular lines in concentric circles, which created an inherently omnidirectional code. That meant it didn't have to be oriented before being pulled across a scanning slot. On the other hand, the concentricity brought with it a redundancy that took up extra space. Zellweger's variant of RCA's code cut out a significant portion of the arc of the circles in order to partially compensate for this weakness.

The information in these bull's-eye-type codes was decoded by comparing the width of the arced line that made up each circle with the width of the space between those circles. Basically, the scanner compared the width of the line with the width of the space.

As it turned out, the bull's-eye produced very high error rates and presented a huge problem in decoding. The reason wasn't obvious at first. But with experience, we recognized that if the black lines that made up the circles were too thick in comparison to the white spaces between them, the reading would be incorrect. That thickness could be caused simply by a printer with too much ink. The too-thick lines also made the white spaces in between them too narrow—a second error. Conversely, if the printer didn't have enough ink, a reverse error would occur. Now, multiply this failure across the whole bull's-eye and the result could be catastrophic.

This misreading was especially problematic in the early days of scanning systems, when coded labels were not part of most package design. The labels had to be printed by inexpensive in-store printers and applied to the packages by store personnel.

The bull's-eye code may not have been reliably readable, but it was proposed by mighty RCA and tested by the company at a Kroger's store in Cincinnati, Ohio, to give it legitimacy.

The bull's-eye code initially created by Joe Woodland and Bernard Silver at Drexel University in 1949 initially had no chance, because the technology for scanning and incorporating it into a system was not available at the time it was invented. However, I believe that, even if it had been selected as the UPC and employed in the 1970s when the technology was available, it did not have the inherent protection from systematic errors or the essential reliability to provide the world with a successful and ubiquitous code.

On that point, the early failure of the bull's-eye barcode was fortuitous, and not just for us. If this circular code had been chosen as the industry standard, the resulting failures would probably have prevented point-of-sale scanning from becoming a success, at least in our time frame. One of the problems of a full or partial bull's-eye code is its circular shape. No matter how a circular code is positioned in a printer in which the paper or the ink move, the direction of the motion will be perpendicular to an edge of some of the lines of the arc or circle that are to be read. This is not the case with a code made up of parallel bars and spaces, since the bars can all be lined up in the direction of the motion. This is critical because motion perpendicular to the long edge of the bar will cause a smear to occur. Smears lead to misreads.

Properly aligned straight bars only smear at the end of the bar which is of no consequence. At the beginning of the program, and for many years thereafter, most barcodes were printed in the store. This meant that low-cost printers were paramount, and low-cost printers had a lot of this type of smear. Large error rates during this early period would have doomed the barcode systems.

The patent rights to Woodland and Silver's bull's-eye code were acquired by RCA. Ironically, more than twenty years later, Woodland—a short, fast-talking, balding man in glasses—was employed by IBM in their Mohansic, New York, facility. He had not been a party to our early work in developing the proposed code and system, but he would eventually join the group in a product planning and promotional role.

In 1967, a different IBM group (based in San Jose, of all places) had built a test system for item marking in supermarkets. This group, used by IBM to develop "custom systems," actually installed working models in the Bay Area. The system used a code that could be read with a handheld wand. It was not omnidirectionally scannable, so it could not be considered a viable candidate for the Ad Hoc Committee's requirements. Still, I did visit one of those Safeway supermarkets and observed the hand wanding. It did work, sort of. But it didn't have the flexibility, function, or low error rate we required in our new system.

LEARNING FROM FAILURE

Numerous other codes, fourteen as I recall, were proposed by as many companies. Even my old neighbor, NCR, had proposed a version featuring a color code of green, black, and white.

In short order, McKinsey and the Ad Hoc Committee reduced the list to seven finalist codes. All these were described in a *Business Week* magazine article on April 7, 1973. The article reported that the seven finalists were Litton, IBM, RCA, Zellweger, Singer, Pitney Bowes, and Charecogn.

Then good news: the article went on to report that the winning code was "patterned closely on the IBM submission." We had won!

An interesting story behind the committee's choice: during a recess of the final judgment meeting, the McKinsey representative told me that our code was clearly the best from the point of view of readability, reliability, and space efficiency. However, he went on to say that the committee felt that if it selected IBM's code outright, it could be accused of being unduly influenced by one of the most powerful companies in the world. So it was decided that the committee could not officially select the IBM code.

My head spun. For a moment it seemed like almost four years of work were now lost. Then I got an idea. Prior to the meeting, George Laurer told me about two minor improvements he came up with for the code. While they were definite improvements, I did not believe they were critical at the time. Moreover, we had already passed the time deadline for making changes to our proposal. But now, I had an idea that might be able to save the day. I told the McKinsey rep that we had made a couple of minor improvements to the code and that we would not claim them as ours. Instead, he could let them "belong" to the Ad Hoc Committee. That would enable it to create and select a "new" code that was not strictly IBM's.

I did not have specific authority from IBM to make this offer. I felt sure the company would have approved the offer if I'd had the time to propose it. But I couldn't be absolutely certain. And for an instant, it passed through my brain that what I had just offered might get me fired. I comforted myself by remembering that, as with all the other companies submitting codes, IBM had previously agreed that any code we submitted would be in the public domain. This was required by the Ad Hoc Committee, since it did not want to give, in its words, a "world record jackpot lottery" to the company that submitted the selected code.

Of course, everybody knew that the company whose code was selected would enjoy a tremendous competitive and financial advantage for having developed the hardware and software to read it. I also knew that if our code was selected, even if not in name and with the minor improvements I offered, we would have the same advantages as if they had selected our exact original offering—namely, all the decoding circuits and electronics we had developed would still read the code.

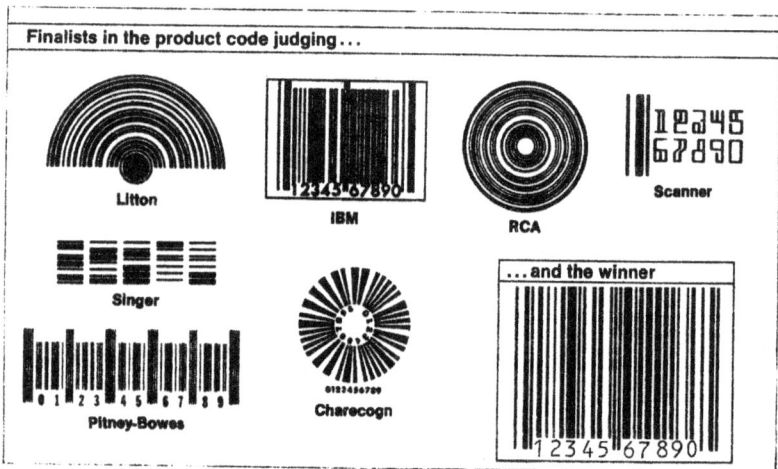

Seven finalists and the winner selected by the Ad Hoc Committee. Notice the winner is the same as the IBM code, except that the top of the bars on the left, center, and right are cropped off.

REGULAR PRINTED SYMBOL DRAWING 6.1

Our proposed UPC symbol (Barcode) with dimensions. Variations to accommodate small or irregularly shaped packages were also proposed.

The McKinsey representative said he would present my offer to the Committee. In what seemed like a lifetime later, he reappeared and announced that the Committee had rejected all the proposed codes and selected their own code. A graphic in the *Business Week* article pictured all the finalist codes, then announced that "the committee rejected proposals of the seven finalists and came up with its own design—one patterned closely on the IBM submission."

Our competition certainly knew the truth: the *Business Week* image showed our symbol with a slight extension of the top of the leading, center, and trailing bars—an insignificant, and essentially meaningless difference. It would have no effect on our reading hardware or software. Back at IBM, I needn't have worried. I received no criticism for overstepping my authority, only congratulations for a great success. I thanked George for getting us across the finish line.

CHAPTER 6

MUCH ADO ABOUT CODES

At the same time as we were designing the supermarket code, we were also working feverishly on a code for the retail industry.

Instead of being an obstacle, this market's need for thirty-six alphanumeric characters proved to be an advantage. The retail code could be read by a wand waved across a retail merchandise ticket. Unlike the supermarket application, it did not need to be omnidirectionally scanned while being pulled across a checkstand at 100 inches per second.

That said, there were some challenges. The merchandising tickets had to be printable in the store or the store's warehouse, then attached to the merchandise. As in the supermarket application, reliability, readability, and space efficiency were important. Almost from the start, we concluded that an optical barcode would not work—it simply wouldn't be able to handle the density and volume of information we needed to convey. So we chose a magnetic code, which had sufficient character density and adequate durability.

As we worked concurrently on both markets, we determined that we could use the same code, or spatial marking, for each. For supermarkets, the code would be optical bars on paper (or other material). And for retail, it would be magnetic "bars" on a magnetic strip similar to the back of a credit card. In retrospect, this was a momentous decision and a huge benefit. It simplified our development process dramatically by allowing both applications to share very similar decoding algorithms and reliability analysis.

As noted, the National Retail Merchants Association, working with the National Bureau of Standards, was the organization that would select the code for the retailing industry. In 1971, two years before the selection of the UPC by the supermarket Ad Hoc Committee and the *Business Week* article (above), I traveled to Dallas for an NRMA meeting and presented a paper describing our proposed code. (The Appendix is a copy of this document.[8]) In the end, our proposed "Delta Distance Code" was accepted as the retail industry standard—albeit not in as dramatic a fashion as for supermarkets.

LESSONS FROM VICTORY

In my view, the Delta Distance Code was the most innovative element of both the retail and supermarket universal product codes (UPCs). Why was it chosen over all the other barcode applications submitted by prestigious corporations? I believe there were four decisive features:

1. The Delta Distance Code was *logic-compatible* with numeric-only and alphanumeric characters.

---

8. The Appendix contains an in-depth discussion of the details we considered in the development of the Delta Distance Code. It also offers technical considerations, and a discussion of the advantages and disadvantages, of several different versions of the code, as well as the mathematical analysis and corresponding equations.

2. *The code was self-clocking,* which meant the operator did not have to worry about pulling the item across the scanner at a constant speed. That allowed for reliable hand-wanding, the primary method in retail and often necessary in supermarkets. This characteristic also permitted the code to work with high reliability even when it had been printed in-store by a low-cost printer—as was the case during the first few years before food manufacturers began printing codes directly on packaging. This was crucial: without that readability and reliability, I don't believe supermarkets would have continued to use the scanners; the costs would have been too great.

3. *The code was a two-level code.* That is, the signal was either black or white, on or off, making it possible to read even when the image was blurred, dirty, or damaged. It could be represented by black-and-white bars on paper or similar material. Alternatively, it could be represented by simple, two-level recording on a magnetic material. This made, and still makes, the code easy to prepare, mark, and decode.

4. *The code used independent stand-alone characters.* In other words, each character was the same wherever it was located, independent of what came before or after. That is, it was not modified by its *context.* This may seem self-evident; but at the time it was very difficult to achieve this constancy. Most competing codes that did not have this characteristic were difficult when using conventional in store printing methods.

An important aspect of Delta Distance C is that it only uses measurements of the leading edge of a bar to the leading edge of the next bar, or the trailing edge of a bar to the trailing edge of the next bar. This eliminates a major problem of systematic errors, such as too much ink coming from a printer. This would cause all bars to be too fat but, since all the bars are too fat, their leading edges move the same amount, and the measurement from one edge to the next remains the same as in the case of a normal amount of ink. In like manner, the trailing to trailing edge measurement also cancels this type of error. Delta Distance A had a similar advantage but was not as efficient as C.

PAIRED CODES

The development of our two codes was a team effort. Soon after we moved the development program from the Los Gatos to the Raleigh Laboratory, I was able to assemble an outstanding team to work on them. George Laurer, Heard Baumeister, and Bill Crouse focused on supermarket, while Jack Jones took retail. I had many years of experience in optical scanning and had spent a bit of time in magnetics, so I acted as technical consultant to both teams.

As noted, both the supermarket Ad Hoc Committee and the retail industry made it very clear from the beginning that all proposed codes from any manufacturers must be "in the public domain." Specifically, we were told not to file any patents on the codes we were presenting. Operating under these guidelines, I instructed our engineers not to file any patents on the code developments.

We were further advised to publish the proposed codes in the public domain (such as trade magazines and appropriate

IBM published documents) so that it would be clear that no manufacturer, including IBM, would ever be able to make ownership claims on any of the codes. In addition, the company whose code was selected was required to publish directions on how to print, magnetize, and read the codes, which, when selected, we did. Manufacturers could file patents according to normal procedures on hardware and software used to prepare and read the codes. These patents presumably do not prevent other manufacturers from coming up with their own proprietary products to read the selected codes.

Retail (by the National Bureau of Standards) and the supermarket industry association had both been advised by their attorneys to instruct submitting companies not to patent their codes. Many years later, this turned out to be a huge mistake when various parties came forward in the supermarket arena to claim that *they* had invented the code, or some portion of it, before we at IBM proposed it. They were utterly bogus claims, but they made life difficult for the attorneys of the Uniform Product Code Council (the body set up by the supermarket industry to manage the codes).

The reason for the difficulty was a lack of documentation. During the patent application and approval processes, detailed files are normally created with information about the inventor's invention. But because no one had applied for patents for our codes, we had no records to prove ownership of intellectual property, nor was such information on file at the United States Patent Office. That dearth of information created space for fraudsters to come forward, claiming that they had invented all or part of the codes. I found out, at the twenty-fifth anniversary celebrations of our invention of the code, held at the Smithsonian Institution, that payments had been made to settle some of these invalid claims. While they weren't (fortunately) in the millions of dollars, they were nevertheless sizable. It would have cost more to go back in history and compile the necessary evidence to defeat these wrongful claims than it would have to simply pay them.

Some of the competing codes were covered by other patents, most notably the bull's-eye code, which was patented by Woodland and Silver in 1949. This, however, was a moot point, since the seventeen years of patent protection had already expired by 1966, and the code was not selected in any case.

In truth, our Delta Distance Codes for supermarket (optical) and retail (magnetic) were not covered by any previous patent. After retail and supermarket systems both became ubiquitous in the marketplace, however, the question was raised: who invented them? Three people claimed to have invented the Delta Distance concept. Two of those people were members of my team, namely George Laurer and Jack Jones. The third was Ernie Nassimbene, an engineer and an associate who had worked with me in the Los Gatos Laboratory throughout the 1960s. Subsequent to our experimenting with Delta Distance A and B, Bill Crouse came up with Delta Distance C, which turned out to be the preferred version.

I believe it is actually possible that each of these brilliant engineers came up with the idea on his own. Some of the early magnetic work in IBM's San Jose Development Laboratory may have used concepts similar to Delta Distance, although that has never been proven. Even sixty years later, as I write this, we still don't know who first came up with the Delta Distance idea. It was quite possibly by a member of our team, and most certainly within IBM.

Here's what I do know: both the supermarket and retail codes contained a host of different elements, developed and incorporated into the codes by the various members of our team. For the supermarket code, I assigned George Laurer to be the lead engineer. He had to select and design many individual characteristics, including the modulus checking schemes, the combinations of bars and spaces that referred to the different numbers, and where and how to place the check bits, to name a few.

I knew that Laurer was especially good at all these things, so his outstanding design came as no surprise. I was particularly impressed by the way he broke the symbol into two separate parts. We could not require the operator to take extra time to orient the symbol as it was pulled over the scanner; he or she had to get an accurate reading in a single swipe, no matter how the code was oriented. This omnidirectional requirement demanded that the symbol be taller than it was wide. This condition is necessary because a scan in the form of an X is guaranteed to pass through all the bars as long as the width of the bars is less than their height. However, if we had taken the entire twelve-digit symbol and made it taller than it was wide, it would have exceeded the Ad Hoc Committee's specification of one-and-a-half square inches of area.

This was Laurer's biggest breakthrough; he came up with a brilliant yet easy way to use different coding (parity bits) for the left and right halves of the symbol. In this way, the scanner could read one half (either half) of the symbol at one moment and the other half an instant later, and then combine the two halves to read the complete number. With this configuration, the symbol needed to be taller than only half its width. Thus, it fit within the required space. Not only that, but this same method would allow simple recognition of a stand-alone code of half the number of digits, which would be required for small packages, or other unique circumstances.

While Laurer certainly contributed so much, he wasn't

alone. He was supported on some of these elements by other engineers, especially Bill Crouse and Heard Baumeister, as well as a physicist, a mathematician, and me. For example, based on my scanning experience, I personally advised him on the resolution capability of state-of-the-art optical technology at the time, which, in turn, defined the minimum size of the bars and spaces. That determined the reliability of the printing and reading of the code.

Moreover, the experts who had coached me in optical scanning during my years in Los Gatos were Mel Rabideau (physicist, optics expert) and Carroll Brown (electrical engineer, scanning expert), both of the Los Gatos Laboratory. Our code mathematician was David Savir, who did a great deal of work calculating the mathematical reliability of the code, which proved its superiority over all other proposed codes.[9]

In short, if our team had been allowed to file patents on the code, I believe that, at least, the names Laurer, Crouse, Jones, Baumeister, in that order, and maybe even Nassimbene, but not Woodland, would have been on multiple code patents.

ANOINTING THE FOUNDER

Let's skip forward a couple of decades, to February 4, 1992, when President George H.W. Bush encountered a supermarket scanner, apparently for the first time.

Briefly, here's what happened: as the country was sinking into a recession, President Bush's handlers decided he needed an upbeat photo-op to improve his position in the upcoming

---

9. "The Effect of the Design of the IBM Proposed UPC Symbol and Code on Scanner Decoding Reliability" by David Savir, International Business Machines Corporation, 1972.

election against Bill Clinton. He arranged to attend and speak at a supermarket convention in Orlando, Florida. After his talk, he was given a demonstration of a supermarket scanning system. When the demonstrator pulled the item over the scanner and then showed him how to do it, the president seemed astonished that the item price immediately came up. Apparently he remarked that he was "amazed." President Bush was reportedly very happy about some positive pictures of him working a scanner, hoping that it might improve his position in the polls.

Just the opposite occurred. The next day, a front-page article in *The New York Times* ran with the headline "Bush Encounters the Supermarket, Amazed." The reporter, Andrew Rosenthal, made the point that the president was so insulated from the typical American's daily experiences that he didn't even know about a technology regular people had been using for more than a decade. Pundits later pronounced this moment as the start of the slide to the president's eventual election loss.

Several years later, Robert J. Kuntz, chairman of the California Engineering Foundation, announced that he had been told the following story by a technical advisor to President Bush: Impressed by his scanning experience that day in Florida, President Bush had asked who had invented the supermarket barcode. The president's staff contacted the Universal Product Code Council and was advised that the barcode had been developed by IBM. The Council had a copy of the original IBM proposal, which listed Joe Woodland as the contact point, since I had assigned him to promote the code and to write our proposal and explanation booklet. The details of what happened after that are not clear, since we do not know what Woodland said. But we do know he was contacted by Bush's people. And we also know that Joe Woodland was given the National Medal of Technology in 1992 by President George H.W. Bush.

For a long time, I didn't even know about this award. By that time, I had left IBM and hadn't worked in the supermarket or retail fields for nearly a decade. When I finally heard about it, I was upset. I wouldn't have included Joe Woodland on a list of team members I considered as the creators of the UPC barcode. The *team* should have received the award, or if only one person could be recognized, then George Laurer would have been more appropriate. But by then it was too late.

With six decades of consideration and perspective on the matter, I would say that Woodland invented the very first barcode specifically intended for supermarkets. Although his code was not used, and for good reasons, he had accurately envisioned the advantages that item identification would bring to the supermarket industry. Some might say that he deserved the medal for just that. Unfortunately, his receipt of the medal has led many to believe that he invented or created the ubiquitous supermarket barcode, or Universal Product Code (UPC), that we see everywhere in the world today. And that is certainly not the case.

Before Woodland was hired to join our team, he had phoned me to ask for my support to transfer him into our CTS marketing organization. He told me that he had studied our vertical bar Delta Distance Code, had concluded that it was far superior to his 1949 bull's-eye code, and wanted to support it and the supermarket program. I thought, *Why not?* Technically, Woodland was the inventor of RCA's competing code. What could be better, I reasoned, than to have him promote *our* code instead?

Given the unique characteristics of the Delta Distance Code, I can't say that the Woodland/Silver code was particularly instructive in any part of the creation of the UPC. Our IBM patent attorney at the time, John Frazoni, concurred: he stated very clearly to me prior to our publishing "The Identification Technology for the Retail Industry" in 1971 (as noted, the Appendix in this book is a copy of this document) that the Woodland and Silver bull's-eye patent was of no consequence to the patentability of the UPC, which is a Delta Distance Code.

Of course, as noted, no such UPC patent was ever applied for. How different official history would have been if patents had been allowed.

CHAPTER 7

ALL IN THE TIMING

On the twenty-fifth anniversary of the universal product code, a recognition ceremony was held at the Smithsonian Institution. Dr. David Allison, the chairman of the Division of Information Technology and Society, Smithsonian's National Museum of American History, interviewed George Laurer, Joe Woodland, and me about the development of the code.

My interview was the last of the three. At the end of our conversation, Dr. Allison commented that, all of a sudden, he understood the answer to a question that had been haunting him for some time: why hadn't an automatic scanning system for supermarkets been successfully introduced much earlier? The answer, he continued, was that the necessary supporting system components and the system itself had not been available nor even envisioned before.

He was right. More specifically, before the helium neon laser, there was no safe light source. Before high-density magnetic contact recording, no file system existed that could handle the massive storage and recall requirements with sufficient speed. And before modern communication systems, the speed and cost of transmitting the data between the scanners and the store storage system or controller would have been too slow and expensive.

READYING FOR REAL LIFE

In 1973 we had won the competition for the standardized codes in supermarkets and retail, two of the most important sectors in the US economy. Winning, however, was only the beginning. We still had a long way to go.

First, we had to launch the code from the lab into the real world, applying it to real-life retail items in real-world situations. Second, we needed to make the technology so reliable, so cost-effective, and so easy to use that it would be embedded everywhere in daily commerce.

It took only a couple years to create the code. It took a couple of decades to achieve these next two goals.

That said, we were way ahead of NCR in that I was scanning items before the code was even selected. Still, IBM's demanding test procedures required us to take a lot longer even before introducing the equipment to a "public" test environment. And that guaranteed we would have stiff competition—computing and engineering corporations around the world would be working hard to catch up with us, developing the supporting systems, technology, and components to make the code actually function in the real world. We had to fight every step of the way.

One of the most important system components we needed to design was the point-of-sale (POS) terminal itself. In this regard, I was most fortunate to have found an absolutely brilliant young engineering manager, J. Leonard Felton.

Len was a graduate of Virginia Polytechnic Institute, but with considerable IBM experience, especially in chip-driven computer processors. He turned out to be an outstanding engineering manager and later a very successful executive. With Roger Kaus, who was part of the earlier Rochester Lab cash-register project, Len formed a great team and implemented a POS terminal of which we would all be forever proud.

Len's team designed a POS terminal consisting of a lightweight box that housed a stored-program computer. This type of computer stores instructions in its memory that lets it perform a variety of tasks in sequence or intermittently. Keep in mind that this was a decade before the arrival of the personal computer (PC). Since our terminal would have its own intelligence, we could write programs that could execute the necessary terminal operations and still provide the flexibility demanded by the many different supermarket and retail chains.

For example, all stores share the need to print the item description and price, and sum up the total on the customer's checkout receipt. But different retail chains may wish to format that list in different ways, or to incorporate unique comments and advertisements into their receipts. This was impossible with current cash registers. But with a stored-program computer in our terminal, we could easily accommodate those unique requirements.

Programming techniques were somewhat primitive in the very early 1970s. Even then however, we could offer some customized features through firmware rather than pure software. "Firmware" was the term used to describe the placement of computer programs in factory-written, read-only memory (ROM). Various customers described to us exactly how they wanted their systems and terminals to function, and we programmed the specifications into the terminals at the factory, using this new ROM firmware technology. This provided economies of scale: almost all our customers wanted the same firmware at all the stores throughout their chains, so we could program thousands of terminals identically. While it wasn't convenient for the customer to change (each terminal would have to be updated individually) the new technology had the advantage of ensuring that every terminal delivered to a particular customer would operate identically, no matter where it was located. We did provide customers with the ability to make minor changes using the terminal's regular read/write memory.

As we anticipated, customers loved this new capability. Point-of-sale now meant greater accuracy and productivity during the checkout process, and the ability to incorporate advertising and marketing.

One of the most important decisions Len had to make was choosing which of several available microprocessors to use in the terminal. According to IBM's rules in those days, we were to give precedence to IBM's own chips (that changed in the 1980s with the Intel 8088 powering the IBM PC).

At the time, various IBM laboratories, especially Kingston, were developing several different microprocessors under the leadership of Bob Miller, who went on to become the youngest lab director in IBM history. Len and some of his team members participated in the design and review of these, which went by names like "the Micro 80." Importantly, the technology had just recently advanced to the point that it allowed the microprocessor to be fabricated on a customized semiconductor chip. This dramatically reduced the manufacturing cost of what was a high-performance device at the time. Further, other IBM development organizations could use the same microprocessor for their applications, so the organization benefited from economies of scale.

At the same time, the firmware and software were implemented in read-only and read/write semiconductor memory respectively, both of which had been recently developed. Inexpensive display technology had also been newly developed. That made it possible for the customer and the checker to see the price and item description in an easily readable form from a good distance, even in broad daylight. Here we used the latest technology of the day, although it was not proprietary to IBM.

The keyboard was easy for us: our Raleigh manufacturing facility had been producing hundreds of thousands of keyboards for terminals used in volume applications such as insurance, banking, and airline ticketing. Adapting that technology for use on the point-of-sale terminal was a no-brainer.

The terminal printer offered a shopper's receipt, a journal tape, and a check-cashing insert station; it was basically off-the-shelf. And because it would get heavy use, we went for the most bulletproof IBM model we could find.

The cash drawer was the sole old-school element in the terminal. We incorporated a simple, single cash drawer for both supermarkets and retailers.

All these components would serve retail and supermarket point-of-sale terminals. They would, of course, require special unique attachments, such as fixed and handheld wand scanners for reading the supermarket code, and a magnetic wand for retail. Finally, both applications required in-store label printers, which we had yet to develop.

THE RIGHT READER

Retail applications could be sufficiently served with just a handheld wand. But we soon realized that supermarkets, which handled many, many more items than, say, a department store, would require handheld wands as well as in-the-checkstand fixed-head scanners. The self-clocking feature of our Delta Distance Code proved to be a terrific advantage in developing handheld wands.

As I noted earlier, the classic retail chains demanded a complete alphanumeric character set with a large number of characters for detailed item descriptions—even the names of fashion designers. That's why we developed a version of the Delta Distance Code in magnetics for the retail application. Interestingly, at the mass-merchandising retailers of the day (for example, Kmart, S.S. Kresge, Walmart, Target (formerly Dayton-Hudson), Zayre, or Caldor) customers brought shopping carts to the cash register loaded with a mix of supermarket and retail items. For these situations, we decided to recommend the optical Delta Distance Code, which required the store management to forego scanning the longer and more detailed alphanumeric information. The faster operation of a fixed-head optical scanner overruled the nicety of the detailed information allowed by the magnetic code. In addition, we knew that, over time, the number of items with factory-marked optical barcodes would increase dramatically.

To get our arms around the vast challenge of the application requirements in these different types of stores, we originally broke down our target markets into four categories: supermarkets, the big three (Sears, Penney's and Wards), mass merchandisers, and traditional retail stores. In the half-century since that time, supermarkets still operate in the same way, using the optical code. The big three (well, those that still exist) have, for the most part, changed from the alphanumeric magnetic code to the optical one. By comparison, the mass-merchandising segment, using optics, dramatically increased its marketplace share through such familiar names as Walmart, Target, and Costco. Traditional retailers used our *magnetic* Delta Distance Code for many decades, but most (as we predicted at the time) are now using the *optical* Delta Distance Code: the version the world now calls the barcode. I don't know of any store that ever used both optical and magnetic codes at the same time.

MAGIC WAND

We considered two different types of handheld scanners. One type was the wand, which looked similar to a pencil and could move along the surface of the code, be it optical or magnetic. The second type of scanner typically had a pistol grip and was pointed at the whole image of an optical code. The pistol-grip type scanner did not work with magnetic codes.

I assigned two engineers to work on handheld scanners. Bill Crouse worked on the pencil type and I worked with Jack Jones on the pistol-grip type. We constructed a model of a checkout line and checkstand in our laboratory, installed a rudimentary scanner in it, and experimented with several types of wands.

Very early in our development program, IBM made a number of changes to our senior management, including changing the president of our division. Replacing Jack Kuehler, the new president was B.O. "Bob" Evans. Bob was a great big heavyset man with an outstanding technical reputation. He was also known for a tough, somewhat flamboyant, management style: a real corporate warrior. At IBM, he was the closest character to Al Shugart we had at the time.

In his well-regarded book, *The Gamesman*,[10] Michael Maccoby analyzes the various styles of technical management. I

---

10. Simon and Schuster, 1976.

believe he would have characterized Bob as a "jungle fighter" with high level "craftsman" skills. Bob's office was in our group's headquarters in Harrison, Westchester County, New York. Soon after his appointment as division president, he visited our lab in Raleigh to evaluate our new point-of-sale system project.

I took him into the lab, where we were setting up the scanner, and told him that we had to be able to pull a coded item across our fixed-head scanner, and read the label while it was moving as fast as 100 inches per second. Moreover, the item might be spinning as it was sliding across the scanner. Not only that, but it might be held as high as six inches above the scanner surface as it was pulled across. Further, we explained, we had to handle the data from as many as forty different scanners in one store at the same time. Bob announced that this was one of the most preposterous ideas he ever heard.

Then he said, "Well, Paul, you do have an outstanding technical reputation, so I'm going to let you work on this for one year. During that year I will not come back, because I know I would kill the development if I did. But in one year, this damn thing better work, because if it doesn't, McEnroe, your desk will be moved directly to the parking lot."

True to his word, Bob left us alone with our full funding for the year. When that year was up, right on schedule, he reappeared at the lab. I took him back to the scanner lab, and he picked up a pack of cigarettes to which we had applied a printed Delta Distance barcode. He literally *threw*, in a sliding, spinning fashion, the pack of cigarettes across the scanner. The scanner read the number and showed the item description on the display.

"Well, I'll be goddamned," Bob said in disbelief. Then with a straight face, he proceeded to remove the cover of the checkstand, as if to confirm that we didn't have a tiny engineer stuffed in there, typing numbers and letters into the display. Bob marched to the parking lot, only to return with three bottles of Jack Daniel's whiskey. We were flabbergasted, because in those days you could easily get fired from IBM for having alcohol of any type in the lab, much less consuming it. From that day forward, Bob was a staunch supporter of our program.

We worked very hard on optical scanning from late 1969 through 1971. In the process, we came up with two working prototype handheld scanners. One was the pencil type, the other the pistol-grip type. As you may recall, the Ad Hoc Committee had advised us *not* to file patents on the code itself. That said, we were still free to patent any scanners we invented to read that code. So Bill Crouse applied for a patent on his outstanding hand-held wand, which took advantage of the self-clocking nature of the Delta Distance Code and was thus less expensive, but required the pointed tip to make direct contact with the actual code on the product. He actually had a version that slipped over your finger like a ring. Jack Jones and I came up with and patented the pistol-grip scanner, which allowed the operator to "shoot" at a barcode several feet away.[11]

_____

11. U.S. Patent #3699312 John E. Jones and Paul V. McEnroe, Code Scanning System, Oct. 17, 1972.

FIG. 1

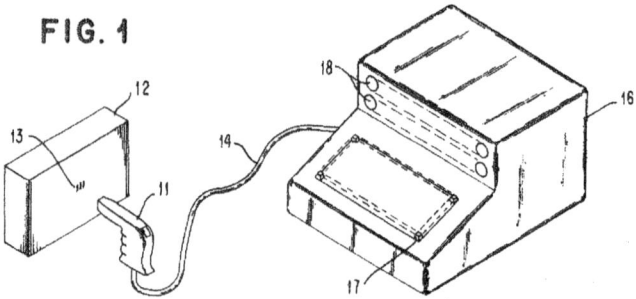

FIG. 2

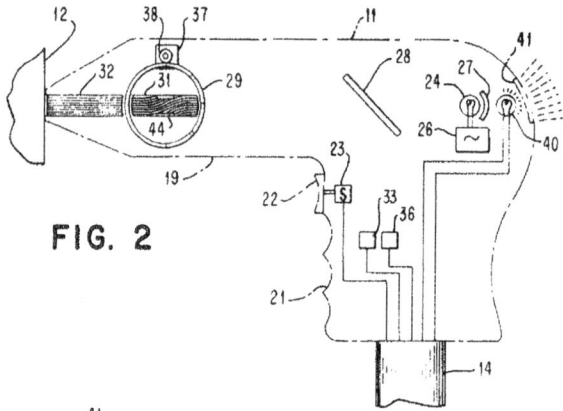

FIG. 3

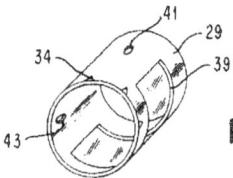

INVENTORS

JOHN E. JONES
PAUL V. MC ENROE

BY *John B Enrone*

ATTORNEY

McEnroe-Jones US Patent #3699312, Code Scanning System, filed Mar. 18, 1971, and issued Oct. 17, 1972. Fig. 1 shows the pistol-grip scanner pointed at a grocery item with the barcode applied. A wire connects the scanner to the point-of-sale terminal. Fig. 2 and Fig. 3 show the internal components of the scanner.

Both scanners worked well. The pistol-grip scanner proved invaluable for checkers, who (as you've no doubt experienced a thousand times) could point-and-shoot across the counter to scan items several feet away that were still in the grocery cart. This type of scanner eventually became the perfect solution for Costco-type applications, in which customers going through checkout typically leave large or heavy items in the cart to be scanned.

As for the fixed-head scanner, the one built into the checkstand itself, we built a working model and installed it in the simulated checkstand in our laboratory. I had spent three of my first years at IBM designing and building optical scanners, so I understood that the light source was among the most important components of our new scanner.

In the early 1960s, I used high-resolution "flying spot" cathode ray tubes (some with fiber optics attached to their phosphor-coated screens), incandescent lamps attached to fiber bundles, and even energy directly from an electron beam. None of these technologies, however, was economically viable for a supermarket scanner. As I researched what technologies other engineers had previously employed for this type of low-cost application, I found that some had actually tried airplane landing lamps in their prototype checkstands—which would no doubt have melted any frozen food items.

DONE WITH MIRRORS

Luckily, since those early experiments, a new type of low-powered laser was developed that was commercially available. Using helium and neon, it was available in the half-milliwatt to one-milliwatt range—strong enough to read a barcode, but not so bright as to blind anyone. The laser produced collimated (tightly focused) light in the long visible-to-infrared frequency range with adequate intensity for high-speed scanning of an optical image.

As the light from such a laser was already collimated, it allowed us to achieve adequate resolution and focus the beam over a large depth of field (as much as six inches), even with an inexpensive lens. (The serendipitous availability of this helium neon laser technology is a classic example of what Dr. Allison of the Smithsonian Museum of American History was referring to when he noted that the supermarket barcode reader couldn't have been invented any sooner than it was.)

Heard Baumeister came up with the idea of moving the beam by oscillating a pair of mirrors in its path. These mirrors, he explained, could be placed in planes orthogonal to each other. One mirror moved the light beam from the laser left to right, while the other took that beam, and in turn, moved it up and down. The speeds of these oscillations were controlled by attaching each of the mirrors to a separate and dedicated tuning fork with a precisely chosen natural vibration frequency.

In turn, the vibration of the tuning forks was synchronized by the electromagnetic signal that drove them. The resultant light beam then moved across a rectangular window in the checkstand from one side to the other and from top to bottom at a high speed. (For those readers with a scientific background or interest, the movement across and up and down follows a sinusoidal motion known as a Lissajous curve. The central part of the pattern is almost linear and forms an X as the cosine part of the Lissajous curve comes back across itself. We designed the system so that only this center part passes through the scanner window. The scan

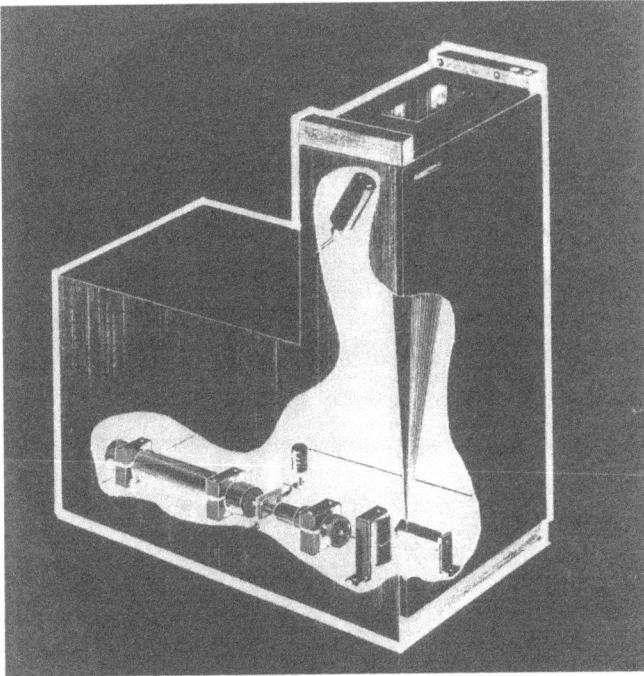

The original checkstand scanner, showing (clockwise from top right) the sidebar safety sensors above and on both sides of the glass window, a photomultiplier tube, the light path going down to the horizontal and then vertical oscillating mirrors at the bottom, the optics and the laser tube.

pattern is enlarged so that the tops and bottoms are blocked out by the sides of the window.)

This *X* pattern of light passes through the scanner's glass window and hits the barcode, which is being pulled across that window. The moving light pattern reflected off the barcode bounces back through the glass and is detected by photomultiplier tubes that have been positioned off to the side below the glass. Solid-state detectors later replaced these rather antiquated tubes.

When we tested our model scanner, and after some fine tuning, it met the requirements of scanning an object that was moving at 100 inches per second, and that was up to six inches above the checkstand window.

We had done it. And unlike the barcode itself, this was patentable IBM technology. No one could copy us—at least not legally.

LASER FOCUS

We couldn't do everything ourselves. And I knew that a very competent group of engineers at the IBM laboratory in Rochester, Minnesota, had considerable experience with printing, scanning, and optical recognition. I negotiated an agreement with Jim Blankenship, the director of the Rochester lab, to have his team build the production prototypes of the fixed-head scanners. The agreement worked out very well: Jim assigned Art Hamburgen to lead the team that successfully delivered a solid, working, manufacturable scanner on time.

As I've noted, the retail application required a simple, handheld, pencil-style wand. The magnetic barcode was recorded on a magnetic stripe placed on the SKU (stock keeping unit) ticket, which in turn was attached to the merchandise. We used the same technology for the ticket that I had used in my previous projects with airline-kiosk and BART tickets: all used magnetic stripes on paper stock, albeit with different codes. A very similar magnetic stripe was just then showing up on credit cards. You may recall my aborted credit card project with Lloyd's of London in the early 1960s. Still I had learned a great deal from that project, and that learning came in handy now. With these experiences, not to mention all the IBM card-handling background available to us, our team did not encounter any serious difficulties in developing our magnetic-stripe ticket printer for the back rooms of warehouses and department stores.

At the same time, Jack Jones developed our pencil-type magnetic wand. The wand was shaped like the letter "L." An operator could simply hold the longer side of the wand, pressing the head at the end of the short side against the magnetic stripe, and pulling it across while maintaining contact with the stripe. Due to the self-clocking nature of the code, the velocity of the pass did not need to be constant. And because the magnitude of the magnetic signal was directly proportional to the speed of the reader across the stripe, we encouraged the operator to move the wand rapidly. When needing to swipe a magnetic striped card, you may recall seeing instructions to "move the card rapidly." Now you know why.

For more technical detail on both the optical and magnetic scanning properties of the Delta Distance Code, please refer to the Appendix. Note that: The best character density or efficiency of the three Delta Distance Codes described in the Appendix is achieved by the Delta Distance C Code. It is also very good for magnetic coding. Delta Distance C effectively employs leading-edge to leading-edge and trailing-edge to trailing-edge measurements, which are much preferred to the more typical bar-width to space-width measurements, especially when scanning lower quality optical barcode printing.

CHAPTER 8

PUTTING IT ALL TOGETHER

We faced another major issue: how to connect the point-of-sale terminals to each other and to the rest of the system.

When an item was scanned, the system needed to transmit that item's code number to the store controller. The controller would then look up the item name and price and/or description, determine how many remained in the store's inventory and, if necessary, reorder more from the warehouse. The controller had to transmit all that information (plus any other special information the store required) back to the terminal and print it out before the customer detected any delay.

But that was just the beginning. This process had to happen simultaneously for all the scanners and/or wands in the store. In a large American supermarket, up to sixteen scanners might be in use at the same time. In the so-called "cash and carry" markets in Europe, there might be as many as forty scanners on the go. In retailing, one store (Macy's New York), had 1,000 point-of-sale terminals. Further, retail required more information to be looked up and transmitted, although it did allow a little more time for information transmission than supermarkets demanded.

These high-speed, look-up-and-reply requirements demanded a lot from two of our system components: the communications line and the file-storage system.

Two of the first engineers I brought into the group had a great deal of experience in communications engineering: Bill Betts and Alex Sawtschenko. Bill was a senior engineer. Alex, while younger, was brilliant at communications technology. They studied the requirements and proposed a "hub go-ahead system," in which the terminals and controller were connected in a closed ring or loop around the store. Different types of devices could then be attached to this ring in any sequence.

In a supermarket application for example, the loop might start in the back room at the store controller, then continue to the front of the store, going from the first point-of-sale unit to the next and so on, then continue to the back room, where it would connect to the manager's terminal and, if necessary, to a store printer. Finally it would return to the store controller.

The beauty of this system was that it transmitted at a low enough frequency to use the telephone lines already installed in most stores. Our system also allowed the lines to run back and forth from a terminal to the store controller and then back to the next terminal, etc., which is referred to as a star connection. This system used existing telephone lines (already installed in a large percentage of the stores) running from the back room to each checkstand. Alternately, if the store was going to run new wire in the process of installing our system and/or renovating the store, the loop or ring could simply be wired from one checkstand to the next and so on, greatly reducing the length of the wires. We believed that this characteristic was a terrific advantage of our system; many, if not all, of our competitors' systems required much more expensive rewiring.

Even though the response to each scan had to be very quick, our system did not require high-frequency communications, because the number of data bits transmitted for each scan was actually quite small. Thus our store loop was very robust, sufficiently quick, and very inexpensive to install. We called it the *Store Loop*, or *S loop*, and finally just *SLOOP*.

A decade later, in the 1980s, as the personal computer began to emerge, the world recognized the need for new, high-performance local area networks (LANs). By then, I was the director of the Raleigh laboratory. In that role, I was responsible for the development of IBM's LANs, and my experience working on networking our barcode readers was the perfect preparation for the task.

CHAPTER 9

IN THE CHIPS

Because our point-of-sale terminal used a stored-program computer (or "engine," as we called it), we also needed interface electronics (adapters) to connect the engine to the various terminal devices. Just a few years earlier, these adapters had been prohibitively expensive to create because they had to be made of discrete components, such as transistors and circuit elements. In a first for IBM, we implemented the terminal adapters with new, specially designed, custom integrated circuits (IC chips). Because these IC chips had not previously been used in IBM commercial products, we had to send our engineers to our semiconductor component development divisions to learn this new technology.

Four decades later, chips can contain billions of circuits. We achieved a density of only several hundred circuits per chip. However, those few hundred circuits resulted in significant cost reductions for the terminals, and we took advantage of that. By the time we were done, we had a custom chip adapter for the SLOOP, POS control, display, keyboard, cash drawer, and miscellaneous devices such as a coin-change machine. Once again, if we'd attempted this a few years earlier, we would not have had the technology to take advantage of this cost-saving opportunity.

LOOKING UP

Up to forty scanners might simultaneously be in action, requesting price, item information, and inventory status—without any noticeable time delay for any of the customers at any of the registers. These specifications far exceeded the capacities of any existing file-storage system.

With this in mind, let me step back several years, to the mid-1960s when I was working in the Los Gatos Laboratory of IBM's Advanced Systems Development Division on a joint project with RAND Corporation. RAND had developed the RAND Tablet, the first device that let users write with a pen or stylus on a flat horizontal tablet and see the drawn lines appear on a cathode-ray-tube display.

The RAND Tablet required a file-storage system that operated at speeds unprecedented at the time. RAND approached IBM to see if we could build a system to meet its needs.

RAND probably came to the Los Gatos Lab because our founder, Reynold "Rey" Johnson, and director, Lou Stevens, were two of the inventors of IBM's first disk-storage system. Rey was known as the "father of the computer disk drive."[12] An elegant, rather formal man, he led the team that in 1956 developed RAMAC (random access method of accounting and control), the first file-storage system that used a disk, rather than tape or drum. That first model recorded five megabytes and weighed one ton. (In 2023 it's possible to buy

---

12. *Los Angeles Times*, "Reynold Johnson; 'Father' of the Disk Drive." Sept. 18, 1998. https://www.latimes.com/archives/la-xpm-1998-sep-18-me-23976-story.html.

several terabytes that weigh about an ounce. One terabyte is a million megabytes.)

Another example of Rey's prowess: while he was teaching science in high school, he worked with two boys who had been assigned to him by the court as part of their sentence for stealing a radio from the school. Rey had them build the world's first electronic scoring machine—just another one of his ninety valuable inventions.

IBM accepted the RAND challenge and put Harold F. (Hal) Martin in charge of the project. (Hal later became general manager of IBM's Rochester, Minnesota, manufacturing and development facilities.) The RAND team was led by packet-switching pioneer, Keith Uncapher, with RAND Tablet creator Tom Ellis as primary engineer. IBM's technical leader was Dr. Joe Ma. I was the program manager. Ma and I reported to Hal. Our team successfully created a solution consisting of a new and quite large high-performance disk with contact-recording heads. (In a 1989 interview describing his career at RAND, Keith Uncapher said that the disk was six feet in diameter. It was certainly large, but not quite that big.)

I use the term "contact recording" because that's what we thought it was until we brought in a college professor and expert consultant, who convinced us that the head was demonstrating "Bernoulli's effect" and actually floating less than a millimeter above the surface of the disk. While the RAND Tablet and our disk did work, together they were quite expensive and never became volume products.

Jumping back to 1969 and the "scanning item look-up" problem, I immediately thought of that high-performance disk and those "contact recording" heads. I approached IBM's current recording experts. The technologies we had developed for RAND, I discovered, had improved to the point where we could now design a device with fixed and movable heads both on a reasonably sized disk. That would solve the high-speed scan look-up problem and also serve supermarkets' normal file-storage system requirements. In particular, the fixed "contact" heads were used for the scan look-up to find the price and item description, while the movable heads served the normal file-storage system requirements.

IBM's primary recording technology location was in San Jose, California. Due to the tremendous growth in file systems, a second laboratory (with great hard-disk-recording expertise) had just been built in our British laboratory in Hursley House, near Winchester, England. Interestingly, the British company, Supermarine, had developed the legendary Spitfire fighter plane in this same building, which looked more like a grand English country house or even a castle than an airplane development facility. Perhaps that's why they chose to develop their fighter plane there, as it didn't resemble a likely target for an enemy to bomb.

San Jose was up to its ears in work, but Hursley's management, led by the sartorially elegant Brit, John Faircloth, accepted our development project. Interestingly, I reported to John for a time when he served as the director of the Raleigh Laboratory, a position previously occupied by Jack Kuehler, and later by me. But John is now better remembered as the chief scientific advisor to England's Prime Minister Margaret Thatcher, a Cabinet office, at which he excelled. He was Knighted in 1990. We created our set of specifications. John and his Hursley team created a special product, code-named "Gulliver," that we could use for both supermarket and retail applications. Once again, no other company had anything like it. I was so excited! Later on, I visited the Hursley lab to personally thank John and his team. (Again, congrats, Sir

His Royal Highness Prince Phillip "checks out" our supermarket checkstand scanner demonstration system.

John.) I also had an opportunity to demonstrate our scanner to His Royal Highness Prince Phillip, who was very impressed and congratulatory.

GULLIVER'S TRAVELS

Every supermarket and retail store required a store controller that managed all the point-of-sale terminals (as well as their attached devices, including scanners and wands), label and ticket printers, management terminals, and the communications network.

Our store controller was much like a computer, except that it was not designed for general-purpose computing, nor did it have attached computer-like input/output devices. But it allowed us to offer custom programming to facilitate individual customer requirements. It used the central processing unit to

IBM Systems Journal

Volume Fourteen | Number One | 1975

The first quarterly issue of the *IBM Systems Journal* in 1975 was dedicated entirely to the Supermarket and Retail Store Systems. All fifteen authors were members of the team.

manage all the above devices, as well as the new "Gulliver" embedded file system. The controller in turn communicated with local or remote host computers, using the standard communication technology and protocols of the era.

The controllers for supermarket and retail were very similar, except that we also offered a failsafe version for supermarkets. This particular unit was a duplexed pair of computer-like controllers tied together such that, if one failed, the other automatically took over and ran the store. This complexity was warranted, because supermarkets typically did not have prices on their merchandise. If the supermarket controller failed, and could not look up prices, the supermarket was temporarily out of operation—an event we could not allow to happen.

We were short on human resources in Raleigh to complete the design and development of this duplexed in-store controller, so I subcontracted the project to my friends in the IBM San Jose Laboratory. They did an outstanding job and delivered the product on time.

For a more detailed account of the design and analysis details of the retail and supermarket systems, including scanning considerations, refer to the *IBM Systems Journal*, Vol. 14, No. 1, 1975. This entire edition was authored by me and my team, and provides an in-depth description of these systems and all the above-discussed system components.

CHAPTER 10

A LIGHT REFLECTION

In 1969 I had launched the Consumer Transaction Systems Engineering development program by hiring six engineers. They were already working for IBM in the Raleigh Laboratory, and I transferred them to this new venture.

For the first year or so, we operated much like a Silicon Valley startup. In addition to my regular job in developmental engineering, I was the systems manager in charge of market planning and business management. Wearing both hats kept me in a near-frenzy; it seemed like almost every week I was either traveling to New York or hosting New York visitors in Raleigh. Fortunately we got a new systems manager during our first year; J. Roger Moody was an outgoing and effervescent IBM computer systems sales manager. After his appointment, I was able to focus all my energy on developing the retail and supermarket products.

My meetings with the New York executives tended to focus on marketing and business issues. Sarkis Zartarian was an expert on these topics—and a workaholic. He spent day and night helping prepare presentations that I made to our senior management. IBM management was just as interested in our direction and progress as any typical Silicon Valley venture capitalists would have been.

A common practice within IBM was to give code names to products still under development so we could easily refer to them in documents and discussions without revealing their nature. Due to my interest in golf, and because the legendary United States Open golf course venue of Pinehurst was nearby, we code-named the supermarket and retail terminals, "Pinehurst." Appropriately, as we subcontracted a great deal of the store controller development to the San Jose Laboratory, we called the store controller, "Pebble Beach."

A TRICK OF THE LIGHT

An entirely unique problem emerged as we developed the supermarket checkstand scanner. We chose a low-power helium neon laser as our light source. When the IBM lawyers got wind of that fact, they balked. "No way are you going to put a laser in that scanner," they told us.

The reason? They worried that it would blind a clerk or shopper.

The lawyers were also concerned about people who had made the mistake of looking at the sun during an eclipse, and who thus damaged their eyes. The lawyers worried that these people might later claim that their "sun damage" was instead caused by the escape of laser light from the scanner. They were concerned that checkstand operators, who might use the scanner on a daily basis for years, would be more likely to make such claims.

From a technical perspective, their fears were not founded. We designed the scanner so the laser light exited the unit to read a barcode only when that barcode was pulled across the window. We incorporated two levels of fail-safe protection to prevent light from escaping the scanner at any other time. None of that mattered to the lawyers.

That said, their concern was understandable. At the time, IBM was constantly being threatened with lawsuits, for almost any reason, mostly by people who wanted to get rich quick. So our lawyers had a valid point. They even created a new term, "laser suicide," to describe people who might intentionally try to damage their eyes by holding open the scanner and staring into the laser in order to be able to sue IBM.

After numerous discussions on the topic, I did some research on my own and found that no existing studies measured the cumulative effect of low-energy laser light on the human eye. We knew what intensity of laser light caused immediate damage, but not about any cumulative consequences.

MONKEY BUSINESS

I decided we needed to answer that question quickly via a source whose reputation was unquestioned. I studied several possible sources and discovered that one of the world's foremost experts on the topic was a researcher at the Stanford Research Institute (SRI) in Menlo Park. I was able to get funding for an SRI project to conduct original research on the long-term effects of Class 1[13] level helium neon laser light on the human eye.

To conduct this research, SRI had to acquire several rhesus monkeys from Africa. The eyes of rhesus monkeys are very similar to those of humans, at least when it came to our concerns. I visited the program several times during the rather

---

13. Of the four classes of lasers, Class 1 is the lowest power. See 21 Code of Federal Regulations (CFR) part 1040 in the United States and IEC 60825 internationally.

lengthy testing process, worried that negative results would likely kill the scanner project. Fortunately, the monkeys were fine. As predicted by the SRI experts, the results showed conclusively that there would be no danger from our scanner's laser light accumulation, even over a period of many years.

Sometime after the laser research was completed, I got a notification from SRI: they were sending all the monkeys to me. Needless to say, the last thing I needed was a group of rhesus monkeys running around the laboratory. I had visions of horror (and I admit, some hilarity) as I thought about what might happen when several crates of rhesus monkeys arrived at IBM's loading dock. I went on an urgent search for a solution to the homeless monkeys' problem. Fortunately, I was able to find, rather quickly, a zoo that was delighted to provide them with a home. Such was not a problem covered in any of my college engineering courses.

LIFE LESSON

By the early 1970s, we built quite a nice testing facility, especially the supermarket scanning test laboratory. We had constructed a supermarket checkstand, complete with point-of-sale terminal and accessories as well as a barcode label printer, and we stocked the lab with a variety of supermarket items to scan. We needed to understand the role of a checkstand clerk and get a hands-on feel for manipulating the groceries in the most efficient manner for scanning. None of us in the engineering team had ever worked as a checkout clerk in a supermarket, not even as teenagers.

Playing "supermarket checkout clerk" was so much fun that my kids sometimes begged me to take them down to the lab on weekends to play with the scanner. They operated the system for hours, pretending they were working in an imaginary supermarket, laughing as they checked out items. I doubt they would have had as much fun doing the job in real life, but they loved playing at being real-life adults with jobs. I still remember their peals of laughter. I didn't realize then that they were among the first to see the world of the future; the world they're living in now.

One of the supermarket test lab's primary objectives was to demonstrate to supermarket executives the effectiveness and speed of the system. The retail system laboratory was set up to demonstrate the printing and hand-wanding of magnetic stripes as a part of the customer-salesperson purchasing procedure with our new point-of-sale unit. While they weren't as revolutionary as our supermarket system, the magnetic SKU tickets were new to everyone.

Looking back, this process was an important lesson: as engineers, we tend to build what *can* be built, not necessarily what *should* be built. Further, we tend to focus upon how things ought to be operated in a perfect setting, and not how they will be used in real-life settings. If companies focused on the latter, the world would not suffer through so many product failures.

ENTER SALES

I've already made clear the importance of a dedicated sales-force to our program: we needed an expert and focused sales organization to prove to customers that the efficiencies of our new system more than justified its cost.

IBM's Data Processing Division was responsible for all product sales, except for typewriters and a few other associated products targeted for smaller businesses. As noted earlier, when we began the project in 1969, John Akers, later to be president and CEO of IBM, was the head of the DPD Distribution Industry sales and marketing portion. In 1972, as our program took on more importance, DPD set up our dedicated marketing organization to manage the sales of just our products, under Marvin Mann, who would report directly to Akers. Marvin was an outstanding person and an efficient manager and one of the most important figures in this story. He was a Southern gentleman whose word was his bond.

Mann set up offices in our building in Raleigh. From there, his sales personnel managed the introduction of our products. This included retail and supermarket demonstrations to prospective customers in our laboratories. After our official product announcements, customers visited these demonstration laboratories almost daily. The scanning laboratory hosted almost all major supermarket chains.

Demonstrations and tests were more important to supermarket than to retail, since scanning was entirely new and crucial to the former. In retail, SKU wanding was also new, but not so different from the previous checkstand procedures. For that reason, we set up two actual in-store grocery tests, one domestic and the other international. Domestically, we set up a full store test in a Pathmark store in Plainfield, New Jersey. Internationally, we set up a testing station in Canada, in a Steinberg's supermarket in the Duval suburb of Montreal, Quebec.

When we began these in-store tests, almost no supermarket items were source-marked with a barcode. Store personnel had to print out barcode labels on our in-store label printer and then affix those labels to products—from Wheaties boxes to cans of Campbell's Soups. Importantly, as discussed earlier, the codes needed to be scannable even when printed by an inexpensive and not very precise printer. This was where our Delta Distance Code made all the difference.

I am convinced that none of the other codes submitted to the Standards Committee would have been sufficiently readable had they been printed with cheap, in-store printers. It took years for grocery manufacturers to integrate barcodes into their source packaging. I'm confident that point-of-sale systems in supermarkets would not have survived several years of in-store barcode printing if scanners weren't able to reliably and accurately read store-printed codes. Fortunately, thanks to the Delta Distance Code, our scanners, and a number of competing scanners, were able to do just that.

In addition to addressing technological problems, we had to worry about any social impact the system might precipitate, and any resulting political concerns that might arise. Once, while visiting the Steinberg store in Montreal, I encountered a Canadian government official also visiting. He wanted to determine whether the new system posed any issues that might be of governmental concern. He was questioning store personnel at all levels, as well as shoppers. In particular, he was concerned about customers' reaction to the fact that prices were no longer marked on each item.

I watched as he stopped an elderly lady who had just finished checking out and was headed for the store exit. Was she upset that there were no prices on the merchandise?

"Oh, no!" she said. "It's actually much better this way, because now I can take my itemized store receipt to the grocery store down the street and see if I paid more or less here."

Previously, she continued, she could not remember exactly what price she had paid for the different items, but now she had a readable list that spelled it out for her. The official was quite surprised by the lady's answer. I never heard any further concerns from the Canadian government.

In the United States however, it was an entirely different story.

IN THE LIMELIGHT

We announced the retail system in September 1973. IBM announcements were generally handled exclusively by the sales organization (DPD). As an exception, I was asked to call on Federated Department Stores as a part of the announcement-day program. The announcement went very well—although it didn't create quite the level of excitement some of our team members had hoped for.

In October, we announced the supermarket system. Again, the announcement went very well, although we didn't break the bank with new orders.

But that was okay. For a development engineer at IBM, the big deal was never the official product announcement, but rather the first customer shipment. In order to have a product approved for an IBM corporate announcement, it was first necessary to prove to an independent technical organization that the product could be manufactured in volume for a stated cost and perform to specifications with excellent reliability.

This independent technical organization was known as IBM Product Test. Meeting all its stipulations was no easy task. For the announcement phase, engineering projections were used to some extent. For the first customer shipment, the entire system had to be built and tested to pass. This was a huge deal at that time for IBM because we had, for many years, been under government scrutiny as a potential monopoly.

The company was terrified of being accused of prematurely announcing products that it might not be able to deliver.

The good news was that we did indeed pass all those independent tests: not only of every component in the retail and supermarket systems, but also of the systems themselves.

THE PRICE OF SUCCESS

This was an entirely new business for us, and only time would tell whether our grocery and retail barcode scanning systems would be successful. Pricing, of course, was a crucial part of determining that success.

We did a great deal of work to figure out the optimal price of each system. Our cost estimates at announcement time supported those price requirements. In the case of supermarkets however, IBM senior executives weren't ready to accept lower levels of profitability. They increased the price shortly before the announcement. We objected, but corporate prevailed; and in the end, they were correct. The increased price ended up generating millions of extra dollars for the company.

My team had taken on an impressive number of development challenges involving hardware, software, and firmware, and we met every one. Of course, some of the work was straightforward engineering design, but a surprisingly significant amount of work required groundbreaking engineering innovation.

Among these innovations were the custom chips—the first chip project of its kind undertaken at IBM, designed jointly by engineers from our Raleigh team and from IBM's labs in Kingston and Fishkill, New York. We also incorporated a completely new magnetic-disk technology, begun in Los Gatos and San Jose, and built by remote team members in the Hursley Laboratory in the UK. We designed the new duplex controller technology with the help of remote team members in the San Jose Laboratory. The fixed-head scanner, pioneered in Los Gatos, continued in Raleigh and was built by another remote team in our Rochester, Minnesota, lab.

All the other components and systems were developed entirely within our Raleigh Laboratory. These included the store communications system (SLOOP); as well as the keyboard, display, and printer; the firmware for the point-of-sale terminal; the firmware and software for the store controller and the manager's terminal; the in-store label printer for supermarket; the in-store magnetic label printer for retail; the handheld wand and pistol-grip scanners; and all the systems considerations for communications and management between stores, warehouses, and customer headquarters. The point-of-sale terminal and the store controllers were developed as different models with functions that varied according to the different needs of individual supermarket and retail applications.

In all, the development of what was now called the barcode was a global effort involving teams of talented people at six different laboratories scattered on two continents.

A CULTURAL CRISIS

In the summer of 1974, we successfully completed our product testing and began regular deliveries of supermarket and retail systems.

Our first regular supermarket installation, not counting the test stores, was at a Giant's supermarket in Tyson's Corner, Virginia. We conducted a thorough test of the system before the scheduled Grand Opening. When the event began, I sat by the phone, waiting to hear from Alex Sawtschenko. I entrusted him to act as the top technical person on-site for this first, crucial, launch.

Well, the phone rang all right. But it was neither the call I was expecting nor the call I wanted to receive.

"The store can't open," Alex told me.

I couldn't believe it. What could have gone wrong? Alex informed me that demonstrators were picketing the store, telling customers not to enter because scanning was going to be used, and prices would not be marked on the items.

I was stunned. I had no idea that such an action might occur at our first store opening. I had worried that some crazy technical glitch might cause a problem. But our technical team was quite capable of dealing with an operational glitch; they weren't prepared, however, for a cultural crisis.

As the days went on, protesters stepped aside, and the store proceeded to operate without any other issues. We were delighted that the scanning system operated perfectly, but we had missed the thrill of our Grand Opening. Moreover, we were now shockingly awake to this new social issue. Later we learned that the picketing may have been organized by labor unions fearful that scanning could hurt employment opportunities for checkstand operators.

Luckily, this was the only situation I know of in which protesters demonstrated over the arrival of barcode technology at a supermarket. As is often the case with a new technology, the public fear was misplaced: with barcode readers, and the resulting increased throughput of customers, supermarkets got bigger and opened more checkout registers. The result? More jobs, not fewer, for checkstand operators, not to mention more technical people to manage the network.

REALITY BITES

By the end of 1974, we had five scanning systems operating in supermarkets in the United States. The systems and all their components worked well. Item marking was coming along, but we still had a long way to go to reach the point where stores no longer needed to print barcodes on in-store printers.

That was the good news. The bad news was that the cultural backlash against our technology continued to grow. It certainly didn't help when popular talk-show host Phil Donahue declared barcodes to be a corporate plot against consumers. In the months that followed, a number of states passed legislation either outlawing the use of scanners completely or allowing scanning systems to be used, but without removing the prices from the items. Of course, without price removals, one of the biggest cost justifications for the scanning system evaporated. By my recollection, eighteen states proposed or enacted legislation banning or restricting the use of scanning systems in supermarkets.

California, for example, required that prices be marked on all items in the store if scanners were installed. Immediately, some supermarkets in the Los Angeles area were found in violation of this law.

What happened? Well, when milk and similar items that needed to stay cold were brought to the store, the delivery person was followed into the store by a local government official. Then, as the delivery person placed the merchandise on the shelves, the governmental official immediately filed a complaint that prices at that store were not marked, specifically those on the items just delivered.

Of course, store management felt it was unfair to be cited

for not marking prices on goods within minutes of their de-livery—especially as these goods needed to be refrigerated immediately. And in fact, stores tried very hard to mark prices in a timely fashion.

The good news for us about the California law was that it was written to expire one year after its enactment, unless the legislature specifically voted to extend it. As people began to accept the scanning system, the objections faded. At the end of the year, the law was allowed to expire.

During this time, I found it necessary to travel to many of the states that were passing these laws. My argument was that the price would still be printed on the signage near the item, so there really wasn't a need to mark the price on *every* item.

I particularly remember a rushed trip to Helena, Mon-tana, to meet with government committee representatives preparing a law to be voted upon by the state legislature. The committee was very interested in the system, how it worked, its safety factors, and the advantages it offered to supermar-kets as well as to consumers—namely, that they should see cost and price reductions from the efficiencies of the system. Legislators were interested in reducing the amount of time consumers spent in the checkout line, and in the cost reduc-tions that resulted from the elimination of marking the price of every item. It helped that there was no record of any check-stand operator losing their job to this new technology.

In Montana, as well as in most other states, once the tech-nical advisors to the legislators understood the details of the system and how it worked, they recommended not passing laws against the scanning systems. Meanwhile, while I had always thought of myself as an engineer and manager, I was now living the life of a lobbyist.

One federal legislator, who shall remain unnamed, was on the verge of preparing a bill for consideration by Congress that would directly outlaw the scanning systems. Rather than the lack of price marking, this legislator was concerned solely about the safety of the laser. He talked about the United States military planning to use lasers to shoot down enemy airplanes and missiles. "How," he asked, "can we possibly consider putting lasers in supermarkets so close to so many consumers, while at the same time we are considering using them as powerful weapons to blow aircraft out of the sky?"

In attempting to contradict these ridiculous notions, which I knew were being considered by small groups around the country, I presented the following argument. The power of the checkstand laser we were using was half a milliwatt. That meant it would take 2,000 of them to produce one watt of power. Supermarkets used many, many lightbulbs to light the store and merchandise. Just *one* of those hundreds of bulbs was typically rated at 60 watts— or 60,000 milliwatts. Each of the 60-watt bulbs in the store burned 12,000 times more energy as all the lasers in the ten scanners that serve a very large store. Not only that, but the federal government had defined four categories of lasers, the scanner laser being in the lowest category and not at all comparable to weaponry, which was rated in the highest category. This argument, especially when coupled with the "monkey test" results, put the laser power concern to bed rather quickly.

In the end, the anti-scanner legislators, who had begun loud and strong, slowly faded away.

EUROPE BOUND

With the US market launched (and legislators mollified), it was time to tackle Europe.

Britain and the rest of Europe made up a very significant portion of the immediately available worldwide market for supermarkets and retailers. Our marketing team was very active in understanding the special requirements of those countries, and made sure we built them into our product.

In the winter of 1976, I took several members of my team on a European trip to better understand those market applications and to tell our potential customers about the scope of our offerings.

In London, we met with executives from Harrods and John Lewis in retail, and Sainsbury's and Tesco in the supermarket space. We also visited Galleries Lafayette in Paris, and traveled on to Germany.

The primary difference between European and American stores were the huge "cash and carry" markets, especially in France and Germany. These stores marketed high-volume goods and sold them in large quantities, what a family might buy to last weeks or longer. They boasted aisle after aisle of racks as much as twenty feet high. (In some ways, they preceded the Costcos and Price Clubs that eventually arose in the United States.) These stores had as many as forty checkstands. Our system was ideal for this layout.

The last several days of our visit to Germany provided an unplanned highlight. We had been on the run every day, with meetings from almost daybreak to well into the night. To reward ourselves for our effort, we took a short excursion to Innsbruck, Austria, where the Winter Olympics were taking place. We went up to the mountain to watch the downhill ski competition where Franz Klammer made his famous gold-medal run.

CHAPTER 11

COMPLETIONS AND INSTALLATIONS

Our development organization had grown significantly and changed quite frequently over the years.

We had started in 1969 with six people. I was the development manager for all technology and engineering. Then with Garrett Fitzgibbons's resignation, I also took on the duties of the systems manager, which included business and product planning. After Roger Moody was installed as systems manager, I continued as the overall development manager.

Later, another experienced engineering manager, Bill Brocker, joined the group in various capacities, including systems manager. At one point, he was systems manager of retail while I was systems manager for supermarkets. Marvin Mann, who had been the manager for our marketing programs in the data-processing division, was moved to the development division and became the overall manager for retail and supermarkets in early 1976. His performance in both roles was absolutely exceptional. He always gave his full support without reservation. The personification of the classic Southern gentleman, Marvin was always a pleasure to work with and for.

By 1976 our team numbered in the hundreds. It had expanded to include quite a large group to write supporting software programs for retail and supermarkets. These groups were managed by their respective application managers, namely Len Felton for retail and me for supermarkets.

AMERICA'S DEPARTMENT STORE

Retail shipments had gone smoothly from the very beginning. By the mid-1970s, they were progressing well when an interesting situation arose with Macy's.

The department was seeking proposals for an entirely new point-of-sale system in its huge New York store—the biggest department store in America—to be followed by similar installations in all other Macy's-owned stores. The store's senior executives called for presentations from the companies submitting proposals for the project.

We were one of those proposing companies. I took Alex Sawtschenko with me to New York to study Macy's existing wiring. We needed to determine whether we had to rewire the entire store to install our system.

As I previously discussed, one of the advantages of our store loop communication system was that it used standard telephone wires. When we examined the flagship store's existing wiring however, we found that no one knew where most of the wires in the telephone system went or whether they were properly terminated. As a result, we knew it would be hugely expensive to rewire a mammoth building like Macy's New York. If another vendor could use the store's existing wiring, its total fee would be much less than ours.

I also knew it would be a nightmare to install a huge, multimillion-dollar system with close to a thousand point-of-sale terminals and have it not work. So I told Macy's we would need to rewire the building.

We did not get the order. A competitive vendor did. They installed their equipment using Macy's existing wiring—and they couldn't make it work. So Macy's returned their equipment, rewired the store, and installed our equipment, not only in the New York flagship, but in all its stores. That was a terrific public relations coup, one that might have instead been a PR disaster if we had tried to force our original "budget" solution.

Meanwhile our supermarket business oscillated between excitedly positive and suffering from political and social backlash. We regularly swung from a large number of orders to a large number of cancellations.

At one point, the IBM board of directors met in Raleigh. This was quite an unusual event, and a very big deal. During the course of my presentation to this group, I was asked quite disapprovingly why I hadn't foreseen all the negative concerns (around price removal, the fear of checkstand operators losing jobs, laser safety) that resulted in so many supermarket order cancellations.

How does one handle a question like that in front of the board of directors of the largest company in the world? Luckily, I didn't have to. IBM's CEO, Frank Cary, jumped up and vehemently announced that that question should be addressed to him. He went on to say that in briefing him earlier, I had pointed out that these problems would likely come up. Furthermore, he said he had personally decided to take the risk and told me to go full speed ahead.

The person who had asked the question sank back in his chair and was not heard from again. I'll never forget Frank

Paul McEnroe (second from left) demonstrates the Supermarket System to IBM President and CEO Frank Cary (third from left) and Hursley Lab Director Sir John Fairclough (far left), Raleigh Lab Director Jim Bookstaver (far right).

Cary putting himself in the middle of the controversy to pull a young engineering manager out of a murky situation. A classy move by a classy man.

THESIS

When I first moved with my family to Raleigh, I had completed all my courses for my engineering degree from Stanford. I had only to complete my research and publish my thesis. But soon I was working day and night, on the other side of the country, on the point-of-sale and barcode ventures. On top of that, I lost my thesis advisor, Dr. Henry Eyring, who had moved to become president of Brigham Young University Idaho and a top leader in the Mormon Church.

Now with my big IBM project fully underway, I found myself yearning to finish my education. So I contacted Stanford to see if it would let me back in. I was pleased when the university agreed. Further, even though I had lost my thesis advisor, Stanford agreed to change my business research topic to point-of-sale.

Of course, preparing a thesis is easier said than done. Deeply bogged down in my exciting engineering and business challenges, raising a family, and wanting to spend a lot more time with Ann and our three children, I had procrastinated about writing the thesis.

Worried that I would never complete the degree, Ann presented me with a proposition I could not refuse. She knew that I was a horse lover—I had just bought a pony for our oldest child, Maureen. She proposed that I could dictate my entire thesis to her, which she could type for publication. Once I finished, I could pick out a horse and buy it for myself. The most amazing part of the bargain? She would take care of the horse for the rest of its life.

Now Ann was not a horse person at all; in fact, she never rode. It was an extraordinary offer, and I took her up on it. Ann typed the thesis, 190 pages of it. I bought a horse. And Ann took care of him for what would be, sadly, the rest of *her* life. We named the horse Stanford, Stan for short. Our children and I enjoyed him for more than thirty years.

My thesis topic, the point-of-sale system/barcode, presented an issue. Stanford expected to own the rights to work done for the thesis, but IBM also expected the rights, and in this case may have been required to hold those rights in order to be able to grant them later to the Supermarket Institute or to place them in the public domain. Stanford, IBM, and I ended up resolving the issue by removing all the item-identification and barcode material from the document. Personally, I thought that the removal ruined the thesis; but the remaining material, on all the other aspects of point-of-sale systems in supermarkets and retailing, was sufficient to secure its approval from Stanford. The conclusion of my thesis was "that almost all the major retailing chains will have point-of-sale systems installed in the majority of their stores by 1980."[14] That conclusion proved to be nearly correct, missing the target by a couple of years.

In June of 1972, after almost twelve years at Stanford (you can certainly call me a slow learner), I finally earned my Degree of Engineer with a Specialty in Business Administration.

AN HONOR

In the years that followed, as the barcode and IBM's point-of-sale technology slowly took over the world, I broke away from the Raleigh Lab from time to time to advance my education as required by IBM's continuing education policy. Accordingly, I undertook a new program in Modern Engineering at UCLA.

In July 1974, I was invited by Leonard Kleinrock, one of the fathers of the Internet, to give a series of lectures at the Weizmann Institute of Science ("the MIT of Israel") in Rehovot, Israel. I had gotten to know Leonard during my studies

---

14. McEnroe, Paul V., *An Analysis of Point-of-Sale Systems, A Stanford University Thesis.* June 1972.

at UCLA. A developer of ARPANET, he was the first person to send a message between two computers on that precursor to the internet.[15] The lectures were part of a conference on communications and networking. There I got my first real look at ARPANET, the germ of the future internet, and a unique early glimpse into the future of my industry, my company, and my own career.

One of the lectures I gave was about the communications system we had created to transmit data from a store's computer and control units to point-of-sale terminals, scanners, manager terminals, and printers located throughout the store or campus of store buildings. It was essentially a new technology and (although the term was not really in use at the time) one of the first examples of a local area network (LAN).

My time in Israel was involuntarily extended as the Turkish invasion of Cyprus temporarily closed the Ben-Gurion Airport, which resulted in an opportunity for me to spend extra time with Leonard. He was not only a mathematical and packet-switching genius, but also a very fun guy with whom to tour Israel, particularly Jerusalem. He knew a surprising number of local cooks and shop owners, Arab and Israeli, who treated us to wonderful dinners and interesting conversations, often speaking with Leonard in their native languages.

The unplanned extension of my stay in Israel gave me the opportunity to visit local supermarkets and retail stores to observe their point-of-sale operations. They closely resembled those I had observed back home and in Europe except

---

15. Britannica.com/biography/Leonard-Kleinrock: On Oct. 29, 1969, Kleinrock and his student sent the first message on ARPANET from UCLA to Stanford Research Institute (SRI), which is regarded as the first message ever sent on the internet.

for one key difference: a lot of people in the supermarket checkout lines carried automatic rifles. They were mostly young men and a few young women of Army age, and in uniform. The Yom Kippur War had ended less than a year earlier, and the rifles reflected the mood of continued apprehension. I can tell you that it is very difficult to focus on the food you are buying or the price you are paying for it while you're looking down the barrel of the automatic rifle hanging over the shoulder of the shopper a few inches in front of you.

The lectures and conferences at the Weizmann Institute were all focused on communications, although that topic encompassed only a small fraction of the varied technologies I had worked on in our point-of-sale developments. Interestingly, six years later, I was asked by IBM to have my team in Raleigh work on a new, general-purpose LAN. I called on Leonard several times to visit Raleigh and consult with my team on our networking developments. In the end, we developed major expansions, as hypothesized in Rehovot, to the store loop (the SLOOP), transforming it into what became an international standard LAN: namely, the Token Ring.

CHAPTER 12

RECOGNITION AND DEPARTURE

Not too long after our first supermarket system shipped, I was contacted by our division president, Bob Evans, and told to attend a big IBM conference in Los Angeles. Apparently, I was going to receive some sort of recognition at the event. I told him I was much too busy to attend, but he wouldn't accept no for an answer. So at the last minute, I grabbed a redeye to LAX.

The event was at the Century Plaza Hotel in Century City, Los Angeles. It was indeed an impressive affair. Three movie screens in panoramic fashion were set up behind the enormous stage. The audience consisted of approximately a thousand IBMers, all looking very spiffy in their classic blue suits, white shirts, and black wingtips. Several of us received IBM excellence awards, all for engineering work.

To my surprise, I received the President's IBM Excellence Award for the creation and development of the supermarket system, including the barcode. The award statement included the words "for the creation and development of the supermarket system from inception through shipment." The award itself, presented by Bob Evans, was a large Steuben crystal bowl set upon four silver eagles, underscored with the words "IBM Excellence." For a young engineer who had come from an orphanage and the backwoods of North Carolina, it felt like an impossible achievement. I was overwhelmed with emotion.

SHARING RECOGNITION

We shipped our first customer retail system in 1974. In the next three years, we witnessed exceptional growth. It was, in all, a pretty solid market entry.

On the supermarket side, the laser-safety and labor-union employment issues had begun to subside. Previously canceled orders were reinstated. Almost all the major supermarket chains had planned to install the scanning systems in almost all their stores. The majority of these orders came to IBM, which was amazing, given that this was an entirely new business stream for us. By 1977, we could see that our venture was going to be successful, and we would be a force to be reckoned with in the barcode field for a long time to come.

We had developed what I believe was the first major "distributed system" in IBM. All our work over the last eight years led me to the conclusion that it was this overall system that enabled the success of the barcode.

I gave as many IBM awards as I could to the very deserving members of my organization. I also recommended several people for major corporate awards beyond my authority to grant. George Laurer received one such award for his work on the barcode. Another went to Joe Woodland for his work on marketing the code and interfacing with the Ad

Hoc Committee. My two other recommendations for major awards went to J. Leonard Felton for his work on developing the point-of-sale terminal for retail and supermarket, and for his overall management of the retail system; and to David Mackie for his contributions to, and management of, the supermarket system. Len and George were at the head of the long list of people who had supported me exceptionally during the eight years of intense work on the development of the retail and supermarket systems—Len in terms of technical management, and George for technological innovations on the barcode and scanner. I must also note Sarkis Zartarian for marketing, and Marvin Mann for management. I can't thank them enough.

The standards organizations originally set up the code selection in such a way that no company or individual could have an active patent or proprietary rights to a chosen code. Of course, companies would make products that operated on the code, and sell those products for profit, but no royalties would ever be paid for the code itself. The personal awards described above were appropriate and well deserved, but no individual got rich on the barcode.

These awards, in many ways, signaled the end of my own time with the Barcode Revolution. I was sad to see it end—but also immensely proud of what we had achieved. In our own small way, we made history and changed the world.

MOVING ON

As it happened, an enticing new opportunity came my way.

In the mid-1970s, IBM had engaged a business psychologist to study the company's most senior executives, including the CEO, to see what personal characteristics they used to run the company. They decided to go a step further and study young people who management thought were on their way up in the organization. The psychologist was Michael Maccoby, himself a student of Erich Fromm and a global expert on leadership and the workplace. He asked me to be one of the subjects in his study, Rorschach test included, on the personalities of successful managers. I agreed, and met with him frequently, occasionally with Ann.

Interestingly, although he never mentioned me by name, Maccoby incorporated several of the phrases I had used in our conversations into *The Gamesman*, the landmark book he wrote on the subject.

In 1977, I was offered a promotion to director of technology and, simultaneously, director of product assurance, for IBM's Systems Communications Division (SCD). The job was in Harrison, New York, where the managing headquarters for the division and the Data Processing Product Group (DPPG) was located. I accepted the positions—and my time at the Raleigh lab came to an end. At least for a while.

At that time, I took some time to reflect on my career and my profession, and on the previous eight years as we created our code and system: from inception and concept, through international standards, and finally volume delivery and market leadership. I believe that the best possible application of engineering is to use science to solve societal problems and create better lives for people. The "cash register problem" I had stumbled onto in 1969 had given me the opportunity to serve untold numbers of people around the world. I felt so lucky!

UP NORTH

The eight years we spent in Raleigh had been very good to me, Ann, and our three children. Now with some trepidation, we purchased a lovely three-bedroom home on three acres in Ridgefield, Connecticut. In Raleigh, we had become a horse-loving family, so the horses moved along with us. The public school system was excellent. The commute was about forty-five minutes each way in good weather; but as I found out later, a couple of hours in the snow.

Both of my new jobs were in IBM's Systems Communications Division. SCD was responsible for all IBM's communications-oriented products, hardware and software. That included all IBM's user terminals, except for typewriters, and a few low-end office systems intended for small businesses. SCD's products also included all the devices and software necessary to connect any product that had to be remotely attached via leased or public telephone lines to any computer system. It included special, industry-oriented systems such as banking and (my previous home) point-of-sale.

All IBM's products were developed in their own technical laboratories around the world. Most of these laboratories were associated with a manufacturing plant, usually located on the same site. The SCD laboratories were located in Raleigh and Charlotte, North Carolina; Kingston, New York; Gaithersburg, Maryland; Hursley, England; La Gaude, France; and Fujisawa, Japan. Each lab director reported to the SCD division president, Allen J. Krowe, who had just replaced Bob Evans.

As director of technology for SCD, I was responsible for ensuring that the development programs in all the labs were based on solid, state-of-the-art technologies that would not cause product or system failures in the future. I needed to be aware of budding technologies and disseminate information about them to labs that might be able to incorporate them in future products. I was also responsible for pulling together the division's product development plans. This was a "staff" rather than a "line" position, to which I was accustomed.

I inherited a small but solid headquarters staff with outstanding expertise. All laboratory engineers were encouraged to raise problems or concerns to my organization. We either provided direct help or referred them to other sources either within the company or outside.

I now traveled frequently to all the domestic, European, and Asian laboratories, often joined by an expert from the staff. Fortunately, no unsurmountable technological problems arose during my year-and-a-half tenure in this position. Much of my time was taken up by helping the business units in each division prepare their product development plans for presentation to the senior executives at the division and group levels.

My second job was as SCD's director of product assurance. The development personnel in each laboratory reported to the local lab director. Notable exceptions were the lab engineers responsible for testing the newly developed products. Those testing engineers were organized separately and reported to their local site's testing manager, who in turn reported to me.

That was key. In my opinion, one of the most important strategies that contributed to IBM's success was that they kept their testing organization completely independent of their development organization. Otherwise, development managers might have cut corners. Further, the testing manager, reporting up through a different management chain, was measured solely on the reliability and performance of the product, and not the timing or the dollar amount of its installation and revenue. The largest laboratories, such as Raleigh and Kingston, had about 150 engineering personnel in their local testing groups, while the others probably averaged about 100. Thus, about a thousand people reported to me in this capacity.

During my early days in this new position, I attended many meetings in which technologists from different labs presented complex strategies to the headquarters staffs, of which I was now a part. I was particularly impressed by the presentations on communication software given by a young manager from the Kingston laboratory: Ellen Hancock. She was cool under argumentative and aggressive questioning from high-level staff and management.

I thought she would be a great choice for the local site product assurance manager's job in the Charlotte development laboratory, which was just being opened. I did, however, have some concerns. The first was that the newly appointed engineering manager in Charlotte was a very, very tough guy, and known to have a bit of a sailor's mouth.

My management was concerned that a refined lady like Hancock might find it difficult to deal with him. The second issue was that her husband also worked for IBM in the Kingston, New York, area. I did not want to pressure them to separate physically, as her potential new job would require a move to Charlotte, North Carolina. Finally, her current job was as a first-level manager of just a few people, while the position I would be offering was as a third-level manager of what would soon be a 150-person organization.

Regarding the first and third concerns, I told everybody, "Just trust me. She'll do fine." Regarding the second one, I worked really hard to find her husband a job in Charlotte at or above his current work level. I had to find it in another project area, as IBM did not allow spouses to report to each other.

I was shocked when Ellen accepted her new role, and her husband did not. Naïvely, I thought this would be a big problem, but she advised me to the contrary. Bottom line, she moved to Charlotte, he stayed in Kingston. I scheduled my standard-assurance meetings only on Mondays and Fridays so that she could come home on the weekends, work a day in the New York headquarters (which was only a couple of hours from Kingston), and then fly to Charlotte. To my surprise, she and her husband often rode their motorcycles back and forth.

Hancock did an amazing job. She handled all the problems, and made significant contributions to our testing methods in complex hardware and software systems. She even handled her foul-mouthed engineering manager. I knew she would go far—and indeed she did. She would become an IBM senior vice president. After leaving IBM, she became COO of National Semiconductor, where her training prepared her well to deal with one of Silicon Valley's toughest CEOs, Charlie Sporck. After that, she faced the ultimate test—dealing with Steve Jobs—as chief technology officer of Apple Computer. Finally, in 2000, she became CEO of the $29 billion Exodus Communications.

ANOTHER LIFE

In the years that followed, I continued to rise through the ranks at IBM. My personal life, however, did not go as smoothly.

In the summer of 1978, I was able to get away for a rare vacation. I decided to attend the World Championship Equestrian Three-Day Event in Lexington, Kentucky. In the process, I injured my back so badly that I could barely walk.

Many of my horsey friends assumed I had fallen from my horse while jumping a challenging fence. In fact, I hurt myself getting luggage from the trunk of my car. On the excruciatingly painful drive back to Connecticut, I had to stop at a roadside rest area in Pennsylvania and crawl from the car to lie flat on my back on an uncovered picnic table during a driving rainstorm, while my kids watched out the car windows in wonderment.

For several months, I was in and out of traction in a special hospital bed in our home. Whenever possible, I moved business meetings to my house. A constant stream of IBM visitors from around the world flowed through my home almost every day. They brought with them presentations painstakingly drawn out on the famous IBM flip charts. One group would be presenting to me in our family room turned hospital room, while another waited their turn in the living room. It actually all worked out quite well. The doctors finally had to resort to surgery. The operation was successful, but the recovery period was quite long, so the meetings in our house continued for another several months.

A HOMECOMING

In 1978, IBM underwent major changes in terms of organization and executive personnel. I was offered a major promotion to Group Director of Systems Development for the Data Processing Product Group. In this capacity, I managed a very large and technically competent staff of approximately 500 engineers and scientists. These experts worked closely with laboratory engineers who needed help solving particularly difficult problems. I reported through Don Gavis to Art Anderson, the group executive. Don was an older, tough,

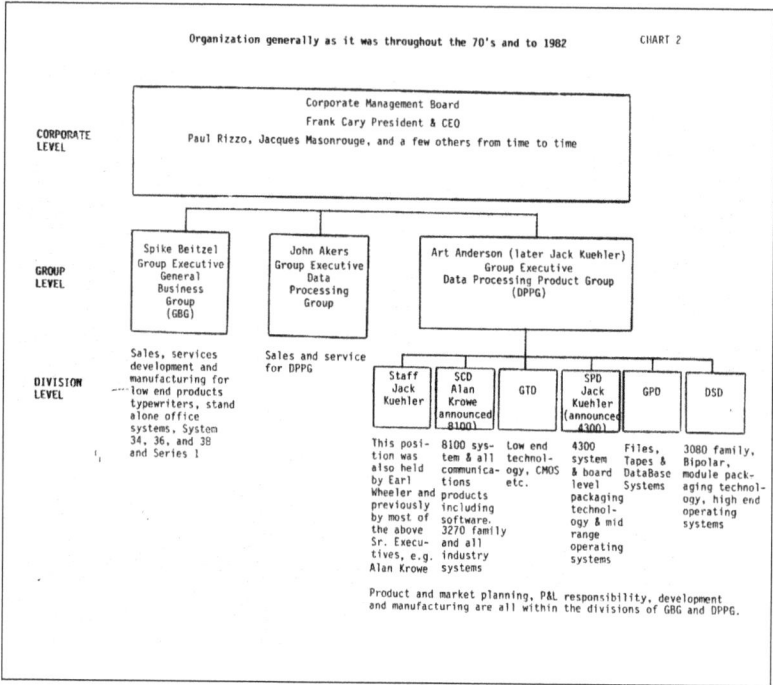

Organization generally as it was throughout the 70's and to 1982   CHART 2

CORPORATE LEVEL

Corporate Management Board
Frank Cary President & CEO
Paul Rizzo, Jacques Masonrouge, and a few others from time to time

GROUP LEVEL

| Spike Beitzel Group Executive General Business Group (GBG) | John Akers Group Executive Data Processing Group | Art Anderson (later Jack Kuehler) Group Executive Data Processing Product Group (DPPG) |

DIVISION LEVEL

Sales, services development and manufacturing for low end products typewriters, stand alone office systems, System 34, 36, and 38 and Series 1

Sales and service for DPPG

| Staff Jack Kuehler | SCD Alan Krowe (announced 8100) | GTD | SPD Jack Kuehler (announced 4300) | GPD | DSD |

This position was also held by Earl Wheeler and previously by most of the above Sr. Executives, e.g. Alan Krowe

8100 system & all communications tions products including software. 3270 family and all industry systems

Low end technology, CMOS etc.

4300 system & board level packaging technology & mid range operating systems

Files, Tapes & DataBase Systems

3080 family, Bipolar, module packaging technology, high end operating systems

Product and market planning, P&L responsibility, development and manufacturing are all within the divisions of GBG and DPPG.

This chart shows an overview of the IBM organization at the Corporate, Group and relevant Division levels during the development period of the barcode systems. Many of these positions and individuals are discussed throughout the book, especially in Chapter 12, Recognition and Departure.

---

experienced guy with a big heart. Art was a brilliant scientist (he had a PhD in physics) in a sea of engineers and business executives. Whenever I explained a complex technical problem or program, he would say, "Sure!" I loved that, because I knew he really understood the points, and in detail, which was not always the case when I presented to top management.

One winter day, I wore cowboy boots to work. I don't remember why: maybe because of the snow, or maybe as a minor rebuff to IBM's famous dress code. After work, I trudged

Left to right, Sir John Fairclough, Hursley Lab Director and Vice President Systems Communications Division (SCD), SCD President Bob Evans and Raleigh Lab Director (later president of IBM) Jack Kuehler.

through the Harrison, New York, parking lot in my boots. As I neared my car, I saw another late-nighter striding through the snow to the only other car left in the lot. It was Art. He too was wearing cowboy boots. His appeared to be made of an exotic snakeskin. In any case, they overshadowed my plain black leather ones; but cowboy boots are cowboy boots, and they were as rare as hen's teeth in the IBM executive lot, even in the snow. Throughout my entire career at IBM, I saw only one other person wearing cowboy boots. That was Don Estridge, the creator of the IBM PC.

The DPPG group included the company's main development, manufacturing, and sales divisions. The products in the group included all IBM's primary computers, file-storage systems, semiconductors, and communications products. The communications products were developed by the Systems Communications Division, which included terminals, front-end processors, and special industry-oriented systems such as point-of-sale and banking. Once again, point-of-sale was back in my life, as was, among about twenty other labs, Raleigh.

One of my responsibilities was to review the proposed engineering product-development plans and strategies of all the divisions in the group, and to present or support the presentation of those plans to IBM's group executive and the corporate management committee, which was made up of the president and two senior vice presidents.

This group director job came with a lot of stress and travel. Every day it seemed, some new problem arose somewhere around the world. The problems were very different, ranging from programming to semiconductors, even to personnel issues. I worried about making even the slightest mistake, since the higher the level of executive title, the more serious the errors could become. We faced several major technical challenges, most notably with a new, multilayer ceramic thermal conductive module (TCM) that was the key to our new, soon-to-be-announced, high-end computer systems. Members of my staff played a key role in solving this problem, which, left unresolved, would have resulted in the company not being able to ship our complete line of new 370 computer systems, causing a multibillion-dollar crisis.

The company was also overwhelmingly concerned about losing our development and manufacturing technologies to Japan. For some reason, I was selected to visit some of the more successful Japanese facilities and meet with several of their industrial leaders. I reported back to corporate that I was very impressed—and not a little concerned—by what I saw.

As I reported to IBM CEO Frank Cary, Japanese industry boasted several qualities that might propel that country to a leadership role in manufacturing over the United States. Those qualities included the dedication of the Japanese workers, science-based quality control, manufacturing and robotic engineering, long-term corporate management commitment, and a corporate and national economic strategy that recognized long-term rather than short-term economic performance. It was a chilling glimpse of the economic battle that was to come.

RETURN TO RALEIGH

In early 1980, the Systems Communications Division president, Allen Krowe, for whom I had worked in my previous headquarters assignment, offered me the directorship of the Raleigh Laboratory.

Technically, the group director job I held at the time was actually a higher-level position than the job I was now being offered. That said, the experience of being a lab director would probably be necessary to attain promotions to even higher-level line positions. I accepted.

After three years in New York, I was going back to North Carolina. I looked forward to it: I had loved raising my family in the Raleigh area, and I enjoyed line management much more than staff. Best of all, I would again assume responsibility for the supermarket and retail programs, which were still centered in Raleigh.

I moved to Raleigh in the spring of 1980 and was joined by my family over the summer. I bought a large brick fixer-upper home with a nice pond on twenty-five acres on the outskirts of Raleigh. It was a perfect spot for our family, complete with cats, dogs, and horses.

THE RALEIGH LABORATORY

The Raleigh Laboratory, also known as the Research Triangle Park Laboratory, had grown modestly since I'd left. It now housed about a thousand engineers, working on the supermarket and retail programs, communications systems and products, communications systems programming, keyboard and display systems, and general engineering support. Over the next four years, I built it up to more than 2,000 engineers. While some of the growth came from transfers, the majority were new college hires. I sent managers and staff to seventy-five universities in what was one of the most significant IBM hiring programs of the time.

SUPERMARKET AND RETAIL SYSTEMS

Obviously, supermarket and retail point-of-sale systems had a special place in my heart. There I made my name at IBM and in the larger technological world. In the three years I had been in New York, both of these systems had enjoyed continued success. In supermarkets, we were just finishing the development of a new holographic scanner, a project begun just before I left. It was the first use of holographic technology in a high-volume consumer product. It did not disappoint. The holographic scanner increased performance in two important ways: faster scanning speeds and depth of focus improvements that added significantly to scanning efficiency.

Our retail and supermarket systems both enjoyed excellent reliability. In response to customer requests, we continued to add performance enhancements, mostly in the form of software, which we used to personalize store operations. Most of these changes were made for a particular chain, such as Giant Foods or Macy's.

Our market share also continued to grow dramatically in both supermarket and retail—enough that we could now safely call our entry into supermarkets and retail stores a success. We had moved from no position in the industry to become its leader.

For the moment at least, we had won the Barcode Revolution.

COMMUNICATIONS SYSTEMS, PART 1: LOCAL AREA NETWORKS

Shortly after arriving back in Raleigh, I received a call from my division president, Allen Krowe. He asked about a new product announcement he had just seen in the newspapers. Its name was Ethernet, and it had just been announced by Intel, Xerox, and Digital (DEC). I told him what I knew about it. He asked what we offered competitively.

"Nothing, really," I told him.

"How many PhDs do you have down there?" he asked.

"A lot," I replied.

"Well," he said, "you'd better get something quick, and it better not look like Ethernet."

At that moment, the development of a new local area network (LAN) became one of IBM's top priorities.

The world was hungry for new technologies to improve communications between terminals and computers within buildings or on a campus. A decade earlier, my own point-of-sale team had developed an in-store communications link called SLOOP, as I have previously discussed. SLOOP was actually a specific LAN, but it lacked features and performance that would allow it to compete as a general-purpose LAN. We had, however, already started adding those features and much higher performance speeds that transformed it into a high-performance, general-purpose LAN. Since it transmitted a leading electrical signal (called a token) over a wire connected sequentially to a loop or ring of numerous terminals, we called it the "Token Ring."

I asked Bill Brocker to manage the Token Ring/LAN development. Bill, who was stocky and tough, and also totally fair, had managed a host of IBM product development programs. I had worked under him from time to time, but now he was reporting to me. Both ways, the experience was good. We were fortunate to have Dan Warmenhoven, a young engineer from Princeton, to undertake the detailed design of the logic and chips, and to interface with Texas Instruments, the manufacturer to whom we would subcontract the chip manufacture. Not only was Dan brilliant at the intricate design, but he was outstanding at presenting his work to any level of executive, whether or not that person had a technical background. I actually tried to get Dan stock options, claiming that if we could keep him, he would one day be an IBM senior vice president, but my efforts fell on deaf ears. The powers-that-be decreed that he was too young.

IBM's Token Ring was introduced in 1984. Soon it was adopted as an international standard by the Institute of Electrical and Electronics Engineers (IEEE). Not long thereafter, Warmenhoven left the company to join Hewlett-Packard. Later he took on what became a most successful role as president and CEO of NetApp.

An interesting article in *TechRepublic*[16] claimed that the Token Ring was technically superior to Ethernet, but that Ethernet eventually won the long-term position in the industry because IBM had overpriced the Token Ring adapters. Either way, Token Ring and Ethernet competed head-to-head for a long time and Token Ring was a successful product for decades before finally bowing to Ethernet.

SWITCHING (CBXS)

In 1980, as I was taking over as director of the Raleigh Lab, the announcement of Ethernet and numerous other factors made it abundantly clear that the commercial world of "in-house" or "on-property" communications was on the cusp of a major change. Transmission of electrical signals, voice, and data across public property would continue to be the jurisdiction of the Federal Communications Commission. Voice and data transmission within a building, or from one building to another on a privately held campus however, was fair game for any company that wanted to offer products.

Two types of technological systems had the potential to become the dominant digital solution to in-house or on-campus voice and data communication. They were LANs

---

16. John Sheesley, "Does ANYONE actually still USE Token Ring?" *TechRepublic*, April 2, 2008. https://www.techrepublic.com/article/does-anyone-actually-still-use-token -ring/.

and computerized branch exchanges (CBXs), the digital version of the old, analog, private branch exchanges (PBXs).

Most of the team in Raleigh, as well as those in IBM research, thought that LANs were the most likely long-term volume solution. Nevertheless, given IBM's position in the field of data management, we could not afford to pick the wrong system to develop.

Therefore we decided to develop both types of systems until it became obvious which one was going to win. I asked Len Felton to lead our version of a CBX type development. Len was a well-built athletic guy, who had to buy the coat jacket several sizes bigger than the trousers to fit properly into an IBM business suit. Len possessed the rare combination of brilliant technological acumen and excellent management skills, along with an exceptional ability to explain complex systems to technical and/or nontechnical customers or senior managers.

As discussed, the LAN program proceeded successfully: we developed the Token Ring. The PBX development project, however, was an entirely different story.

As Len proceeded with the systems design, we quickly concluded that we needed help in designing the actual telephones that would work with the computerized control unit. That unit would in turn prescribe the switching of voice and data between the various digital telephones. For this project, we turned to Motorola, which had a great deal of experience in designing personal headsets for voice communications (remember the Motorola flip phone?).

The work with Motorola went very well, to the point that they began setting up a new operation in Austin, Texas, to develop and subsequently produce the handsets for our system. In the midst of all this, we got some shocking news. One of

IBM's senior executives, our old friend Jack Kuehler, had separately and independently decided that IBM should acquire at least a portion of the leading company in the CBX world in order to ensure IBM's future position in that field. Jack wasn't aware of the program that Len and I were operating in Raleigh.

How did this happen? First, our Raleigh program was brand-new. Second, Jack was a very senior corporate executive not related at the time to the Raleigh lab. He was also on the board of trustees at Santa Clara University, where he met with Robert Maxwell, one of the founders of the telecom company ROLM. The relationship between ROLM and IBM blossomed very rapidly. The minute I found out about it, I contacted our corporate office to inform them that our development program was in direct conflict with the ROLM/IBM negotiations.

It was quickly decided that someone needed to get on the next plane to the Motorola headquarters in Schaumburg, Illinois, to apologize for the situation, halt any further work, and ask Motorola what we needed to do to keep our joint relationship in good order. I was elected to make that call.

I was pretty darned nervous on the flight to Motorola's headquarters to meet with Robert Galvin, the CEO and the founder's son. I opened the meeting by reiterating the above story, detailing our Raleigh development program in cooperation with Motorola and the newly discovered situation between ROLM and IBM. Mr. Galvin was totally understanding, disappointed, but kind beyond words. I asked what we could do to make up for what had happened. He said that there would be no consequences, since we had come and told him the whole story immediately upon our own understanding of the situation. Ever since, I have held Mr. Galvin and

Motorola in the highest possible position of respect. He was a gentleman of pure class.

At that point, we terminated all the work we had been doing with Motorola and began discussing what we could do to support the future of the IBM/ROLM venture. Len met with Robert Maxwell as the program between the two companies progressed. In May 1982, IBM purchased 15 percent of ROLM, and acquired the entire company in 1984.[17]

COMMUNICATION SYSTEMS PROGRAMMING

In my search for a programming director, I turned again to Ellen Hancock, who accepted my offer and, as expected, performed brilliantly. The Raleigh Laboratory continued to be responsible for IBM's front-end processor programming, which included the Network Control Program (NCP), the Telecommunications Access Method (TCAM) and, under the new Systems Network Architecture (SNA), the Virtual Telecommunications Access Method (VTAM). These types of control programs are extremely difficult to write, and even harder to test.

The testing is made difficult because of the immense number of options or different programming circumstances under which the programs must operate. As a result, the Raleigh programming test facility looked like a football field crammed with mainframes. The success of the operation is indicated by the estimate that, by the turn of the century, SNA systems were in use by more than 3,500 companies

17. IBM sold half of ROLM to Siemens in 1989 and completed the sale of ROLM in its entirety in 1992.

incorporating 11,000 IBM mainframes worldwide.[18] I was extremely proud to have Ellen working for me. At the time, the media referred to her as the highest-placed woman in technical management in the United States.

KEYBOARD, DISPLAY SYSTEMS, AND GENERAL ENGINEERING AND SUPPORT

Keyboard development was an interesting issue when I came back to the Raleigh lab. Initially the development went smoothly, and our newly designed keyboard made it into test on schedule.

Then came a crisis. One weekend, I had taken my kids to King's Dominion Amusement Park in Virginia. There I received a call that the keyboard had undergone a hard failure in product test. I immediately grabbed the kids from the amusement park's bumper cars and raced back home to Raleigh, and then to the test lab to investigate.

Numerous keyboards had been taken from the manufacturing line to a test bench in the same room, and several had failed. In particular, a keystroke failed to work. The failing key was different in each keyboard. The failed units were then taken to a different machine for more testing, but we could not repeat the failure. This pattern continued. During an all-nighter, I noticed that the failed units were transported from the test bench to the second tester on a four-wheeled lab push cart. The wheels were hard rubber. I asked to have the units picked up and carried to the second tester by hand. When we did that, the failures did repeat. That led to the conclusion that tiny, dust-like particles must have got between

---

18. "AT&T Outlines VPN Migration Plan." *InformationWeek.* May 12, 1999.

the bottom of the capacitance coupled key and the surface below it, and changed the coupling, thus causing an error. When the keyboard was rolled along on the cart, the vibration moved the particles so that they were no longer under the key. We improved the cleanliness of the manufacturing facility, and asked the employees in that area to wear white coats, emphasizing a mentality of exceptional cleanliness. That was the end of the problem.

In use in the field, this IBM Model F keyboard proved to have incredible reliability, namely a mean time between failure (MTBF) of more than 100 million keypresses. The Model F was the only IBM-designed-and-manufactured subsystem of the company's landmark PC, released in 1980. We were asked to produce several thousand of these keyboards in the PC's first year. In the face of unexpected demand however, we delivered about a million units.

AFTERMATH

Just a year later, a tragedy sent my life spinning in a very different direction. My son, Paul Jr., was killed in an accident.

It was the worst period of my life. Making it even more unbearable were the endless expressions of sympathy from our well-meaning friends, workmates, and neighbors. I could bury my grief in my work. But for Ann, the possibility of meeting all these well-wishers at the supermarket and other locations in her daily life began to take its toll. Equally painful for both of us was seeing Paul's friends growing up and moving on with their lives without him.

We were seeing a wonderful counselor and psychologist, Dr. Jack McCall, to deal with our grief. He had watched us disintegrate. He told us that the best thing we could do was to move away from all these painful reminders. That very night, after dinner, the phone rang and the person on the other end said, "This is Gene Amdahl."

Amdahl invited me to come help him with the management of his new company, Trilogy, in California. Trilogy's president, he explained, had passed away. As CEO, Amdahl desperately needed help in both corporate and engineering management.

This call came completely out of the blue; I hadn't talked to anyone about leaving IBM. What were the odds, I wondered, that I would get a call like this only a few hours after Jack McCall advised me to leave the company to go? Jack understood that IBM was offering me advanced positions in its New York headquarters area, but advised against taking any of those positions, since our three recent years there were heavily focused on our children. Our boys had played on the same soccer team; they were only eighteen months apart. The soccer families were our best friends. McCall said that returning to that area would be a continuing nightmare for our family. Further, the only other available jobs within IBM that would have been an advancement for me would be abroad, and Ann had said she did not want to leave the country. The call, and its timing, seemed to have been written in the stars. I called my good friend Jack Kuehler the next day and discussed the situation with him. He reiterated the IBM opportunities abroad, but concluded that if he were in the same situation, he would probably go join Gene. Trilogy was a high risk, he noted, but Silicon Valley would always offer lots of opportunities that would not require me to force Ann into another move.

In 1984, I retired from IBM, and our shattered family moved back to California and Silicon Valley.

CHAPTER 13

ENDINGS AND BEGINNINGS

Trilogy was, reportedly, the largest Silicon Valley startup to date. Before founding the company, Gene Amdahl had been IBM's most famous computer scientist; he had invented the IBM 360/370 mainframe computer system. Amdahl, a slow-speaking, very thoughtful man, had already founded Amdahl Corporation, the hugely successful Silicon Valley company. It had taken on IBM and won.

Now with approximately $270 million, a huge amount of funding for the era (and surprisingly with no venture money), he had set out to revolutionize the computing world: he aimed to be the first to use wafer scale integration (WSI) to put a massive commercial computer on giant semiconductor chips. Consensus on the street was that he just might pull off this wildly ambitious goal.

I figured that even if Trilogy didn't make it, having it on my resume would set me up again in the Valley. As it turned out, I was right on both counts: Trilogy's aims were indeed too ambitious, and the company eventually imploded. Although the supercomputer didn't make it, I was able to negotiate a merger/purchase of the company with Ken Olsen and Digital

Equipment Corporation (DEC). Digital was attracted to Trilogy's copper polyimide packaging technology, which would lower the costs and enhance the performance of its new 9000 series computer. I led the remainder of the Trilogy team in successfully developing the packaging. Unfortunately, for a host of other reasons, Digital's 9000 was not economically successful. The company did, however, spin out the copper polyimide technology, which has been used successfully in several other high-density and performance products. In fact, the Smithsonian chose to display this amazing packaging technology on the National Mall in Washington, DC.

Trilogy's impressive headquarters, long abandoned, was eventually torn down to make way for Apple's spaceship headquarters.

A SECOND TRAGEDY, AND A THIRD

When we returned to Silicon Valley, we purchased a house in Los Gatos. Ann had always wanted to keep busy and support our local community. After losing Paul, she was even more driven. She found a job as the administrator of a local church and school. Finally after all she had been through, she seemed happy.

And then tragedy struck again. In 1988, Ann was diagnosed with cancer. I have no doubt that her years of grieving had played a part. But she fought the illness like a true warrior—and finally, just before Christmas of 1989, we got the wonderful news that her cancer was in remission.

But it wasn't to be. About a month later, in January 1990, I was sitting in my office at Trilogy when the call came in. It was from the hospital. Ann had been killed instantly in a head-on collision.

It seemed impossible. The drive from our house to her school was just three miles, and Ann was a slow and conscientious driver. I learned only later that a seventeen-year-old girl on her way to school had skidded, overcompensated, and shot across the road into oncoming traffic, hitting Ann head-on.

My wife of almost thirty years was gone.

A NEW BEGINNING

It was time to move on and leave the ghosts behind.

At this point, I was running the Trilogy copper polyimide packaging operation, with Trilogy's outstanding original engineers, and in the Trilogy facilities—only now the company was owned by DEC. Back in 1986, when DEC bought Trilogy, my personal agreement with Ken Olsen was a handshake to the effect that I would stick with the operation until we were able to deliver the packaging technology. I agreed to do my best to keep the most critical resource, our engineers, on board and motivated. DEC knew that the facilities and patents would have no chance of yielding a successful product without the brilliant engineers who developed the technology.

I honored my commitment to Ken, delivering functional and manufacturable packaging, as specified, to DEC. We did not lose any critical employees during the several years it took to complete the technology. Unfortunately, as mentioned earlier, other problems in development and the marketplace plagued the 9000, which turned out to be a financial disaster. Nevertheless, I took some satisfaction in the fact that I had successfully completed my commitment to Ken. I retired from Digital to begin a new life.

My daughter Maureen was the catalyst for that next phase of my life. Maureen was working on her PhD in history at UC Santa Barbara (with, appropriately, a minor in the history of technology). She was also helping me train horses and had made friends with two sisters attending a riding program nearby. At a gathering, she met their beautiful and vivacious mother, Tina. Maureen, who was living daily with my loneliness, saw an opportunity and arranged for us to meet.

Tina and I were married two years later. We decided to move to the central coast of California, buy a ranch, and pursue another great love: raising horses. We found the perfect place in the Santa Ynez Valley, wonderful land, but with no buildings or paved roads.

I had promised Tina the home of her dreams; so as soon as the barn was finished, we spent the next two years designing and building a beautiful classic Mediterranean style home— only to have it burn to the ground just as it neared completion. Our insurance company, the third-largest in California, went belly up at about this time for other reasons. As a result, the fire turned out not only to be a significant emotional loss, but also a financial one. We had already built a barn, so with no other alternative, we moved into the second story of the barn, above the horses.

Fortunately, I had personally put a lot of effort into the design of the barn. I had previously designed our barns in Raleigh, Ridgefield, and Los Gatos. I had used a mortice and tendon design with almost no nails for the Ridgefield barn, which was named the "Barn of the Year" by *Practical Horseman* magazine.[19] Thirty years later it was still featured in their *Handbook of Barn Designs*.

---

19. *Practical Horseman* is the leading publication for English equestrians in the United States.

Now for our new ranch, we were fortunate to find a large volume of reclaimed lumber, including giant original-growth timbers from a nineteenth-century customs house on the Columbia River in Washington State. We had used these amazing materials to construct all our new ranch structures, including the barn and home. While the very best of the beams went up in smoke with the house, the barn was not damaged. Building on my experience in Ridgefield, I designed the Santa Ynez barn as a large, classic, two-story late-nineteenth-century style structure.

More than twenty-five years have passed, and we still live in the barn. Tina may have lost her dream house, but through her prowess as an interior designer, she has turned the barn into a combination of an 1870s ranching museum and an inviting and warm family home. We love it.

We continue to raise horses and cattle, focus on our children and grandchildren, and involve ourselves deeply in local nonprofit organizations. My daughter Maureen, a Fulbright scholar, is a director of corporate development, mergers, and acquisitions, for NetApp, and my son-in-law, Jim Howard, is a materials engineer who worked for Lockheed on the heat shield tile for the space shuttle. My son Mark is a client director in high technology sales for Intervision, and my daughter-in-law Erika is an attorney and senior vice president at Synopsys, all in Silicon Valley. Mark and Erika have two children: Katie is a student at the University of Texas at Austin and Carter is a student at Cal Poly SLO. Tina has two daughters, Tricia and Tara, and a son, Trevor, all with families in California.

HONORS

In 2016 Tina and I were both awarded honorary doctorates from California Polytechnic State University (Cal Poly), San Luis Obispo.

Tina's Doctor of Humane Letters was given in recognition of the vision and leadership she exhibited throughout her career in education. Her notable accomplishments included founding the McEnroe Reading and Language Arts Clinic at the University of California at Santa Barbara. Not only did she found the clinic, she also took it from one teacher (herself) and a few students to a nationally recognized institution serving hundreds of underprivileged students while advancing the art of teacher education in UCSB's Gevirtz Graduate School of Education. Concurrently, Tina saved the oldest one-room schoolhouse in Santa Barbara County from being bulldozed. She had it moved to our ranch, authentically restored it, and now personally teaches "living history days" to classes of local schoolchildren during the school year, and economically disadvantaged children in the summer.

My Doctor of Science was for my technical development career, including the barcode.

The day I received my honorary degree, before a stadium filled with thousands of graduates and parents, as well as the day's speaker, Leon Panetta, Tina surprised me by including these words in her remarks:

I am so proud of my husband, and am in a rather unique position today to tell you a story—Paul's remarkable life story, the more personal side, the part he did not tell you, which is inspiration in itself that few know.

At birth, he was given up for adoption in West Virginia,

154

where his father worked as a coal miner, and as a result, could not support a child. Paul survived an orphanage for two years before being adopted by loving, minimally educated and humble parents who raised him. It was said that they chose Paul from the others because "he did not cry," as some of the more sensitive children in orphanages learned very quickly, if they were to get attention.

But Paul was one of the lucky ones. During his simple upbringing in Dayton, Ohio, he received what is most important in life—*abundant love* and an *excellent education*. Working from a very early age, he ran and managed paper routes, and excelled in high school, earning scholarships to the most prestigious universities, but decided to attend a local college close to home where he could be close to his dying father.

A self-made man who is one of the hardest and most focused workers I have ever met, Paul has persevered and survived not only significant personal tragedies and health challenges, but has also reached pinnacles of seldom-matched triumphs and tremendous achievements. He is a survivor against all odds, a fearless leader, and a brilliant, gifted, and thoughtful role model to us all in what it means to truly persevere.

I have received many honors in my life. I cherish Tina's words the most.

A WORLD ENCODED

As I was contemplating this chapter, I took a break to go to town and put gas in my car. As I pulled up to the station, I watched other customers walking out of the attached convenience store. And I was reminded that inside, every item on every shelf—sunglasses, candy bars, soft drinks in the refrigerators, cans of oil, bags of nuts and snacks, and a thousand other things, all bore a tiny barcode, most printed on their packaging at plants thousands of miles away.

Across the street too, at the supermarket, were hundreds of thousands of individual items bearing barcodes, even most of the fresh food items, including the bagged vegetables and individual apples. The car I was driving, as well as all the other cars around me and on the road nearby, were accumulations of hundreds of parts, all them having likely once been inventoried and warehoused through the use of barcodes.

So too had been my clothes, and the materials that constructed my house, and the bottles that held the medications I took each day, the tools in my garage, and the blankets on my bed. On the television, a series of two-dimensional barcodes on commercials linked viewers to websites for more information.

Letting my mind wander, I could see the world around me, stripped of its substance, leaving only a cloud of the billions and billions of barcodes that made it function so efficiently.

It had all happened in the course of a lifetime—my lifetime. And I had been there at the beginning.

We set out to solve a problem—and in the end, to our amazement, we changed the world.

APPENDIX

The following is IBM's original proposal, authored by Paul McEnroe and Jack Jones, for the Delta Distance Code to be used in the retail industry, as it was delivered to the National Retail Merchants Association in October 1971.

For several years, the National Retail Merchants Association (NRMA) has been investigating the use of Identification Standards. IBM and other manufacturers of business equipment were asked to participate by working with the NRMA's working committees. The objective of these committees is to investigate the technologies that could be used as standards for merchandise and personal identification during the 1970s.

We are delighted to cooperate with the NRMA, because we feel that appropriate standards would be profitable for the retailer and would, therefore, accelerate the use of data processing in the industry.

To respond to the NRMA committees, we have carefully reviewed various coding technologies. As a result of this investigation, we feel that the Delta Distance Code is best suited to the needs of the retailer.

This paper will describe this code. It is respectfully submitted to help you, the retailer, to develop a standard that will be profitable to you and your industry.

P. V. McEnroe

J. E. Jones

October, 1971

Highlights

Delta Distance Code has the following important characteristics: It is both optical and magnetic, alphabetic and numeric and self-clocking. It is represented by two levels, and is a stand-alone (character-by-character) code.

Optical and magnetic

The code can be utilized for either magnetic or optically encoded tickets or credit cards. The retailer, for example, might use printed tickets and also accept magnetic stripe credit cards. Although the tickets and cards are read by two separate devices, the point-of-sale device would interpret the data through common circuitry.

Alphabetic and numeric

The logic of the code allows the use of alphabetic characters. For example, the customer name could be read from the credit card and printed on the salescheck.

Self-clocking

Delta Distance Code is self-clocking. It lends itself to variations in reading speeds as the operator passes the hand-held reader over the merchandise ticket or credit card at the point of sale.

It is also adaptable to a fixed-head reading device. It means that hand-held scanners can be used at the point of sale and batch readers of varying speeds can be used in the stores or at a central location. It is expected that batch readers will be required while retailers convert to point-of-sale systems which utilize automatic entry from merchandise tickets and credit cards at the point of sale.

It allows preparation of readable code by various devices ranging from simple, low-density machines to more costly, high-performance, high-precision encoders or printers.

Represented by two levels

The Delta Distance Code has a simple logic. It has two states such as on or off, bit or no-bit, black or white. It can be represented by black and white bars on paper or conventional two-level recording on magnetic stripes. This is important since it lends itself to simple preparation devices. In the optical form, this means the use of conventional black and white printers of many types.

Stand-alone (Character-by-character)

Delta Distance Code uses stand-alone characters. That is, the representation of a given character is always the same and does not depend on the character before or after. Many codes do not have this characteristic and do not lend themselves to using conventional character printing methods.

Retail systems of the future

The retail store of the future will evolve from the day of the stand-alone cash register, the handset ticketmaker, the batch process computer and the telephone credit checker to a complete, integrated store control system. The system will not only control these functions, but will integrate and interrelate the information to provide tools which will assist the retailer in much more effective management of his store.

The familiar point of sale is only the tip of the iceberg. The system must also control the "back-room" functions such as payments, purchase order entry, receipts, personnel and payroll records, administrative messages, adjustments, returns, transfers, and merchandise ticket and credit card preparation. In addition, it must provide accurate and complete information, on demand, to enable management to obtain a timely picture of his operation, and it must be interactive to the extent that management can modify and control these functions. Most important, it must interrelate the information from purchase orders, receipts and sales to provide a complete and accurate inventory control system.

A key factor in this system is a technology for merchandise and customer identification to identify:

Who is entering the system—customers, employees or vendors.

What is entering the system—merchandise or items.

Progress in retailing in the next decade will require expanded performance from credit cards and merchandise tickets and, perhaps more important, a greater level of reliance by both the *retailer* and the *customer*.

In the area of credit cards, the system must accept a machine-readable card at the point of sale and validate the card, check purchase limits, detect stolen and counterfeit cards and authorize the sale with little perceptible delay to the sales clerk and the customer. In addition, the system should provide for issuance of the retail credit card by the retailer within his own establishment with reasonably simple equipment controlled by the same data base as that used for the normal credit files.

In the area of merchandise tickets, the system must be able to create tickets "in-store" at the point where goods are actually marked and it must read these tickets at the point of sale. The reader should read not only tickets removed from the merchandise, but also coded merchandise information marked at the source. The preparation equipment should provide a range of speeds and costs so that "on-line" ticket preparation becomes economic in applications ranging from the high-volume central receiving center to the remote point in a branch receiving room serviced by one operator part time.

Most important, the total cost of preparing and reading credit cards and merchandise tickets, including the cost of the media itself, should be accomplished at a cost comparable to today's cost of tickets and cards.

Merchandise and credit identification requirements

This new generation of merchandise and credit identification apparatus must meet certain specific requirements. Among these are:

Low Cost - The total system must result in a reasonable cost for all elements including the readers, the preparation devices and the credit card and merchandise ticket media. A major part of the total system's cost is that of the merchandise ticket stock and attendant preparation cost. This is an area where "tenths-of-cents" per unit represent significant cost elements.

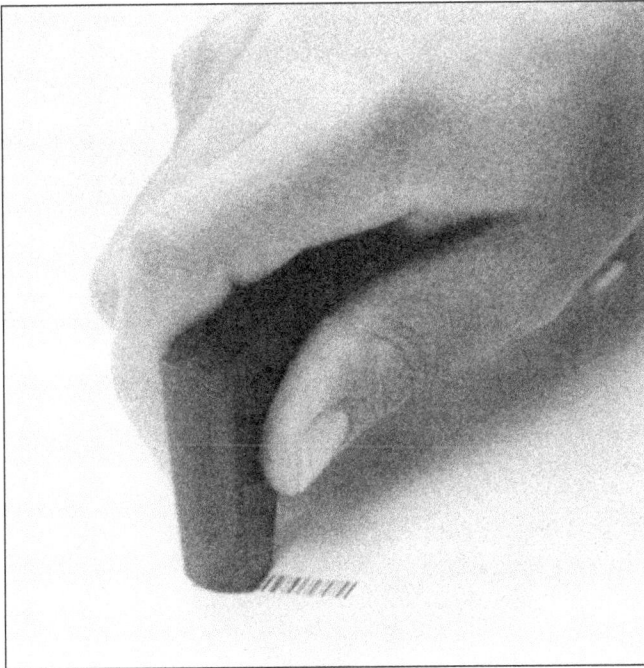

Wand Readers - The system should lend itself to hand-held, operator-actuated "wands" or scanners which are reliable and simple to use and do not require the ticket to be removed from the merchandise.

Multiple Media - The system should lend itself to a variety of coded media including tickets, labels, packages, and credit cards.

Local Preparation - The selected media should be producible in the store on reasonably simple equipment. Some technologies such as holograms, while often fairly simple to read, require rather exotic preparation equipment.

Contrary to the classical approach in the past, the code rather than the *device* is the key to the problem of meeting the requirements of the retail system of the future. The code largely determines such things as the type of reader, the type of media, and the type of preparation device. This paper presents a coding technique known as "Delta Distance Code" which meets the requirements of this system of the future.

Delta Distance Code

Delta Distance Code is a bar code, but a bar code unlike most of the codes with which you may be familiar. Delta Distance Code can be read with a handheld scanner which is actuated in a natural way by the operator by passing the device across the bars. This means that the code not only conveys the information to the system through the scanner, but does so under the conditions of speed changes and the natural arc of the hand as the operator moves the scanning device.

Delta Distance Code is a two-level code. This simply means that it can be printed with conventional types of black and white printers, or alternately, can be recorded by conventional magnetic recording equipment. This makes it unnecessary to invent completely new kinds of ticket and card preparation devices such as those required by more complex codes such as holograms and multicolor codes.

The concept of Delta Distance Code is basically very simple.

As with most codes, a binary "bit" ("1" or "0") must be generated by some difference within the code. Most codes generate this difference by the distance between transitions or by a change of color or some other information. In a code read by hand scanning, the distance between bars cannot be measured conventionally since the operator may vary her reading speed over a range of 10 to 1 or more. Delta Distance Code introduces a novel way to measure this distance; namely, by a timing method which compares the relative distance between the adjacent intervals within the code. This timing method might be called "bootstrap timing" since it picks the timing or clock from within the code itself. This is also known as a "self-clocking" code.

Using this method of timing, Delta Distance Code defines a "bit" as shown in Figure 1. The logic simply compares the time the operator takes to pass across an interval within the code with the time taken to pass across the previous interval over which she has just moved the scanner. The logic compares these times and if the times are approximately equal, the logic defines the second interval as being a "1". If the second interval is not equal, then the logic defines the second interval as being a "0". Obviously, there must be safety bands or separation between the "approximately equal" and "not equal" conditions to account for printer tolerances and operator speed variations. The latter half of this paper is devoted to a mathematical analysis of this problem and a capability assessment of Delta Distance Code when prepared on conventional printers or recorders using today's technology.

As shown later, the code is capable of being prepared at densities of better than 15 numeric characters per inch with conventional printing technology and at higher densities on magnetic recording devices.

The detection or decoding of Delta Distance Code is relatively simple, since basically only two storage devices or registers are required to compare the adjacent intervals due to the "bootstrap" nature of the code. The operator may take only a few hundred microseconds to pass across the intervals within the code; however, counting and storage at this rate is easily within the capability of today's digital logic.

Characters are constructed from a series of bits, using an extension of the concept shown in Figure 1, making each interval a function of the previous interval depending on the value of the bit. Some versions of the code use bars and spaces which vary in width to convey the information, but the basic timing scheme is still the same. For character-by-character printers,

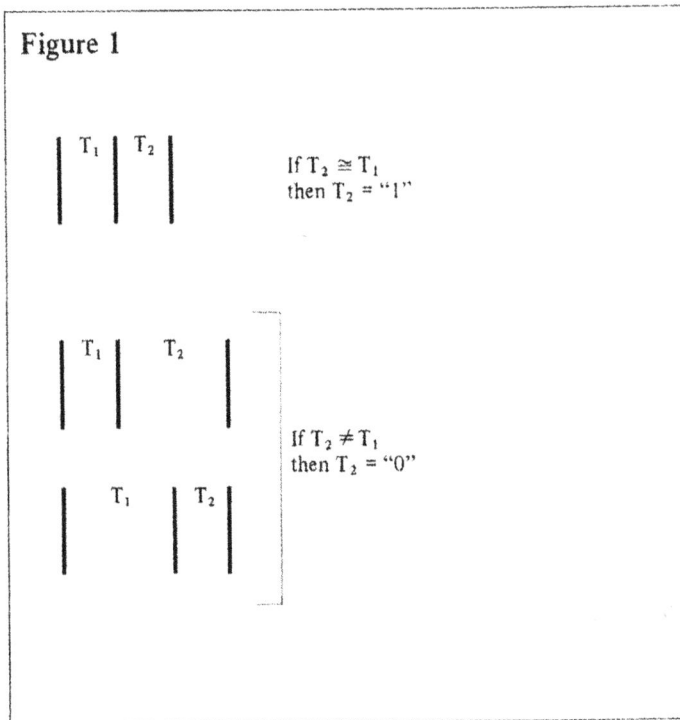

Figure 1

If $T_2 \cong T_1$
then $T_2 = $ "1"

If $T_2 \neq T_1$
then $T_2 = $ "0"

the code can be made "stand-alone" by starting the character with a timing mark and ending with a space which separates the characters but does not carry information. This is the best version of the code for printers such as typewriters or line printers which fall into the "character-by-character" category.

The code can be either printed on paper or recorded on magnetic stripes, thus providing the capability of having a system compatible with either or both of these media.

The remainder of this paper will compare the Delta Distance Code to other types of codes which might be used for merchandise and customer identification and will present a mathematical assessment of the Delta Distance Code, given the variables of printers, recorders, scanners, and finally, operators.

Codes

A large variety of codes exist which might meet the requirements of merchandise and credit identification; however, most were designed for rather specific applications. The early codes were generally designed for data transmission equipment or for recording media such as magnetic tape. These codes are, in some cases, unsuitable for such preparation devices as character-by-character printers.

The character-by-character printer such as a computer driven line printer appears to be one of the applicable devices for credit card and merchandise ticket preparation. Not only is this technology well defined and understood, but the printer also has the capability to prepare both the human-readable characters (name and price, for example) and also simultaneously prepare the machine-readable code on the same pass through the device without requiring an additional recording station. In addition, bar-by-bar printers and continuous magnetic stripe encoders will be used in many applications.

With these requirements in mind, we can list the criteria for the code structure:

1. The code should be *printable* or *recordable*, meaning that it can be printed on conventional character printers or recorded on magnetic media.

2. The code should have the ability to *stand alone* on a character basis. This means that the character must begin independently of how the previous character ended or end independently of the next character. Many codes do not meet this requirement, simply because they were designed for continuous recording devices.

Figure 2: Two-level Bar Codes

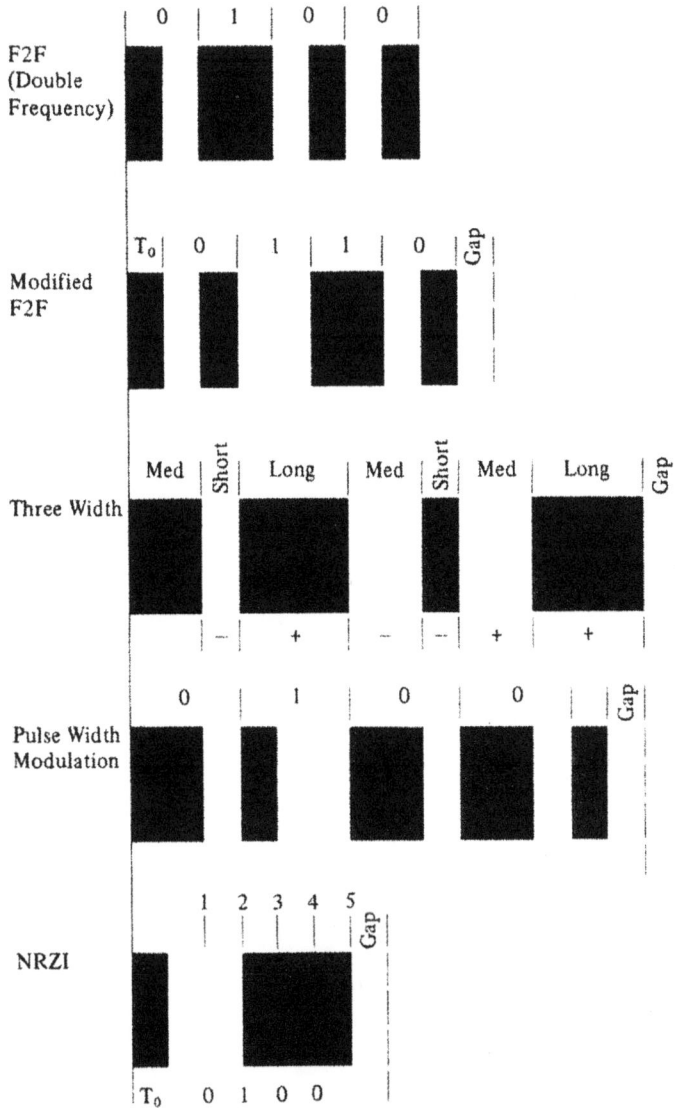

3. The code should be *self-clocking*, meaning that the code can be read independently of the normal velocity variations and accelerations with which an operator will usually move a hand-held wand or other reading device.

4. The code should lend itself to variable density. This means that the code can be prepared on various devices at different character-per-inch densities and is, for example, independent of the specific aperture size of the reader or the particular escapement of the printer.

Codes examined

A number of codes that might potentially meet these criteria fall into three broad categories:

1. Codes which can be printed in the form of black-and-white bars or recorded on magnetic tape. These are known as two-level bar codes.

2. Multilevel printed bar codes in which three or more colors are used to convey information.

3. Codes which are both human and machine-readable, such as those used today on bank checks.

Two-level Barcodes

Several different printable and recordable bar codes are defined in Figure 2. Some of these, such as F2F and NRZI, will be recognized as simply printed versions of codes commonly used on magnetic tape or disks. Others, such as "Three Width" and "Pulse Width Modulation" are probably unique to printed bar codes.

It should be recognized that all of these codes are also applicable to magnetic recording if the bars are interpreted as flux changes on the magnetic media.

The characteristics of these codes are:

- *F2F* - This is the familiar "double frequency" recording in which a double frequency (2F) within a bit space indicates logic "0" and single frequency (1F) indicates logic "1". In the printed representation, a minimum size bar and space indicates "0" and a double size bar or space indicates "1." This code meets the requirements of being self-clocking, since an external clock can be synchronized with the basic bar frequency at the beginning of the message and then be automatically adjusted as the operator moves the wand scanner across the message. The basic problem with this code is that it does not "stand alone" on a character basis and is, therefore, not generally applicable to character printers such as typewriters or line printers. This can be seen, for example, by noting that a character might end with a space or bar which means that the next character must start with either a bar or space and is therefore not independent.

- *Modified F2F* - The "non-independence" of F2F can be rectified as shown by the second code which is called "Modified F2F." In this representation, a "start" or timing bar is added to the beginning of the character and a separating gap is placed at the end. This gives the F2F code character independence, but with decreased character density.

- *Three Width* - This code carries information by virtue of the direction of change in width between adjacent bars and spaces. For example, a change from a medium width bar to a short space is a minus change in width, and from a medium space to a long bar is a plus change. The character illustrated consists of four bars and three spaces of three different widths: long, medium, and short. If it is further stipulated that the character always contains three plus changes and three negative changes, a combinatorial analysis will show that a character set of 14 combinations results.

- *Pulse Width Modulation* - PWM is similar to F2F except that the width of a bar is always compared to the width of the adjacent space. A wide bar and narrow space indicate logic "0" and a narrow bar and wide space indicate logic "1". This code is of constant length.

- *Non-Return to Zero* - NRZI is a semi-self-clocking code which is widely used on fixed velocity devices such as disk drives. A transition indicates logic "1" and no transition indicates logic "0". Even though the clock time is reestablished at the beginning of each character, an analysis indicates that operator acceleration with reasonable character densities results in a high error potential for this code when used with hand-held reader devices.

Multilevel bar code

The multilevel bar code carries information as a function of the transition from one level to the next rather than in the width of the bars. This transition logic for the printed code is shown in Figure 3, which indicates, for example, that if Color A is read first and a transition is made to Color B, then this indicates logic "0". If the transition is made to Color C from Color A, then a logic "1" is indicated.

The major advantage of this code is its efficiency or information carrying ability for a given bar width, but its major disadvantage is that it requires two-color (2 on 1) printing which is generally more costly than single-color printing and requires close registration between adjacent color bars.

Human and machine-readable fonts

There have been various character sets defined which are both human and machine-readable as shown in Figure 4. These range from the magnetic ink characters used in the

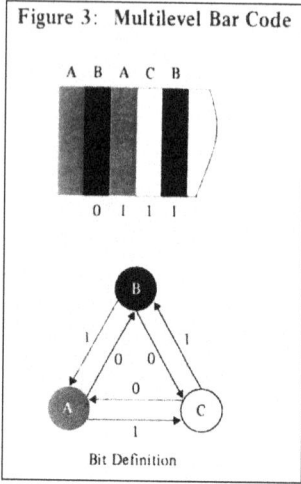

Figure 3: Multilevel Bar Code

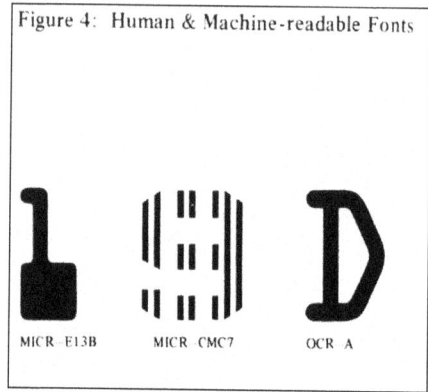

Figure 4: Human & Machine-readable Fonts

banking industry both in the United States (E13B) and in Europe (CMC7) to less constrained characters such as OCR-A designed for optical character recognition systems.

Generally speaking, these fonts are not applicable to hand-held, non-oriented readers. They must generally be read with fixed velocity transports which contain multitrack readers that partition the character into a matrix of vertical and horizontal spaces. The fonts which are less constrained generally require more logic.

The machine-readable character density is, of course, also limited to the human-readable character density. For practical applications, this is generally limited to the range of 10 to 12 characters per inch.

The Delta Distance family

The Delta Distance Code is a family of two-level codes invented specifically for use as a printed bar code, but which is also recordable, and meets the other criteria established above. Versions of the Delta Distance Code are shown in Figure 5.

Delta Distance A (DD/A) - DD/A carries information as a function of the space between adjacent bars. As the operator passes the reader across the code, the time from first bar to second is clocked into a register and compared with the time from the second bar to the third. If the second interval is greater than or less than the first interval, then the second interval is defined as logic "0". If the second interval is approximately equal to the first interval, then the second interval is defined as logic "1".

Delta Distance B (DD/8) - DD/B is similar to DD/A, but carries information as a function of the width of alternating bars and spaces rather than the spacing between adjacent bars. To determine the bit value, a bar width is compared to the previous space width, or a space width is compared to the previous bar width. If the width is different, the bit value is equal to "0" and if approximately the same, the bit value is equal to "1" in a similar fashion to DD/A. DD/B requires the detection of both edges of the bar by the reader rather than the detection of only one edge as with DD/A.

Delta Distance C (DD/C) - DD/C is similar to DD/B, but is read differently and is modified to provide a constant width reference within the character (designated as T_R in Figure 5). The character may be decoded by reading bar widths in the same fashion

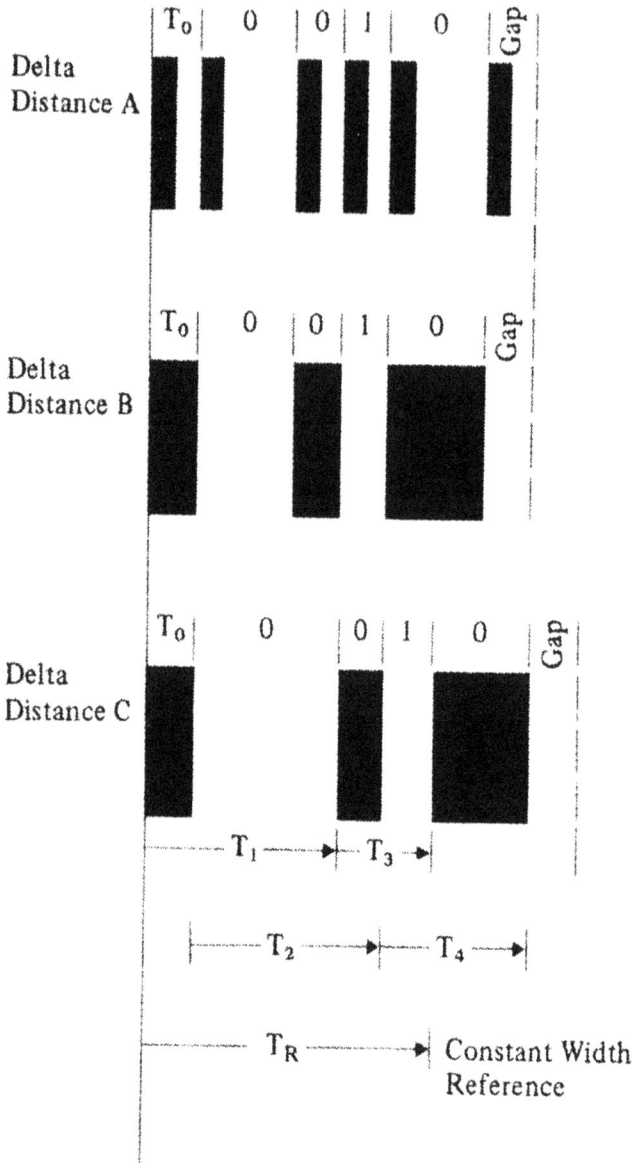

Figure 5

as DD/B, but may also be decoded by reading "edge-to-edge" and storing the resultant four time bases (T_1 - T_4) in four independent registers. Since most impact printers inherently print the edge-to-edge dimension of adjacent bars much more accurately than the width of a single bar, this method of reading results in a higher character density.

The constant width reference (T_R) is used by the decoding device to determine the length of T_1 - T_4. The length of the time bases expressed as long, medium, and short can then be used via an algorithm in the logic to reconvert to the bit value of the character.

Certain types of printers such as those which move in a continuous motion and print individual bars might use a slightly different representation of Delta Distance B and Delta Distance C. In this case, the intercharacter gap can be eliminated and the character would not "stand alone" in the usual sense.

Evaluation of codes and characters

To arrive at a common denominator for evaluating all possible codes for the retail application is a difficult job at best. Two major things do, however, stand out as primary requirements in the retail establishment. First, local preparation of such media as the merchandise ticket at multiple points within the retail establishment is desirable. This implies a reasonably simple, reliable method of preparation. Secondly, the optimum reader appears to be a reasonably simple hand-held "wand" scanner which can read the media with highest character density at multiple point-of-sale locations.

The conventional type of single-color impact printer appears to hold a significant advantage over other methods of preparation such as multicolor printing. The impact printer can, for example, print both the human and machine-readable information on the credit cards and merchandise tickets in one pass through the same device. The technology is also widely used and understood and many existing printing devices can be converted to bar code printers.

On the other hand, the multicolor bar code requires the accurate registration of two colors produced by two different printing actions. If a two-station machine is used, this means that the bars of one color must be printed in the first station and then the document must be accurately moved to the next station where bars of the second color are printed. Problems can occur in this type of printing due to tolerances between print stations which can produce "slivers" or overlap between adjacent colors. Either of these conditions introduces "noise" into the reading system which can result in character error.

By contrast, the two-level bar code character can be printed with a single printing element and single printing action.

In addition to printed bar code characters, there are applications which might require the use of magnetically encoded characters on continuous magnetic stripes. Magnetic encoding might be dictated by character density requirements or environmental considerations, for example. The selected code should be applicable to this type of encoding as well.

The constrained font characters (OCR and MICR) do not appear to lend themselves to low-cost, hand-held scanners within the limits of present day technology and, since their density is inherently limited to "human-readable" density, then it appears that this method of coding would severely limit retail applications at least within the next decade.

A code which can be printed with a single-color printer

or be recorded with conventional magnetics would appear to be the optimum code within the restraints of the application.

Two-level bar code characterization

The primary characteristic in the evaluation of printed two-level bar codes is its efficiency or information handling ability as a function of a given printer or reader capability.

The sections that follow will provide a method of analysis to determine the character density capability of the two-level code.

The character shown in Figure 6 is intended to convey a generalized picture of a typical bar code which contains dimensional characteristics common to most two-level bar codes.

As defined in Figure 6, these characteristics are:

- *Bit* - A bit in this type of code is generally a bar and/or space which may or may not vary in width.

- *Character* - A character is composed of a collection of bits. It might be four bits for a numeric character, six bits for an alphameric character, etc., and may also include an extra bit for initial timing.

- *Minimum Bar Size* - This defines the minimum-width bar used to describe the code and is an important criterion in printable bar codes since it specifies the minimum bar size which the printer must be capable of printing reliably. The minimum bar size is a function of the "character-per-inch" density of the character for a given code definition.

- *Time Slot* - This describes the number of minimum bars and spaces necessary to define a bit or character. The time slot is independent of character density and is an important criterion for judging the relative efficiency of various codes in terms of information-carrying ability for a given printer capability.

- *Character Length* - This is the length of a character from left edge of first bar to right edge of last bar and does not include the gap necessary to separate adjacent characters.

- *Intercharacter Gap* - This is the gap necessary to separate adjacent characters and is generally equal to the maximum printer character-to-character tolerance plus one time slot. The additional time slot is required to provide minimum character spacing if the character-to-character tolerance closes to a minimum.

Figure 6: Character Dimensions

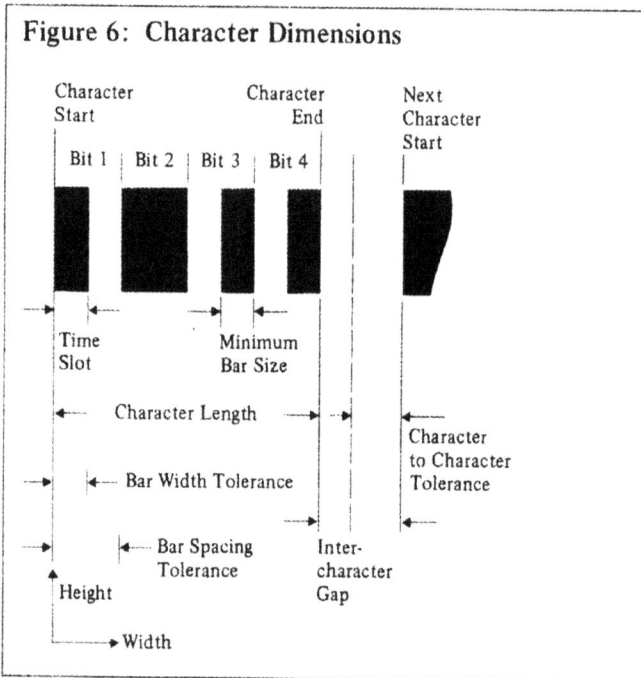

Code	Time Slots per Numeric Character		Type of Reader Required
Modified F2F	10	Notes: 1,2	Width
Three Width	15	1,3	Width
Pulse Width Mod	14	1,2	Width
Delta Distance A	16	1,4	Edge-Edge
Delta Distance B	8	1,2	Width
Delta Distance C	9	1,2	Edge-Edge

Figure 7

Notes:

1 Time slots for maximum length character are shown where characters vary in length.

2 Assumes 2:1 ratios between min/max bar/space combinations.

3 Assumes 3:2:1 ratios between long-med-short combinations.

4 Assumes 3:1 ratios between min bar and max space.

Two-level Barcode Comparison

Some of the two-level bar codes shown in Figure 2 do not meet the parameters set forth in the coding criteria. For example, the F2F code does not meet the criterion that the character be "stand-alone" or character independent. This is because the start of one F2F character depends on the end of the previous

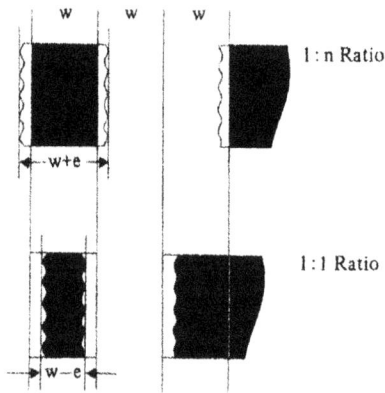

Figure 8: Worst Case Dimensional Tolerances
Delta Distance B Code

$$\frac{w + e}{nw - e} = \frac{w - e}{w + e} \ : \ \text{Crossover Condition}$$

In which: $w = $ Time Slot
$n = $ Long to Short Ratio
$e = $ Dimensional Error

This reduces to: $w = \left(\dfrac{n + 3}{n - 1}\right) e$ 1

The Character Length is: $L = (4 + 2n)\,w + 1$ 2

In which: $L = $ Length
$1 = $ Character-to-character Tolerance

character. Modified F2F does, however, meet the stand-alone requirement.

Likewise, NRZI is not a true self-clocking code because a time base established at the beginning of the character must be used as the "clock" throughout the character. This type of character is subjected to errors due to operator velocity

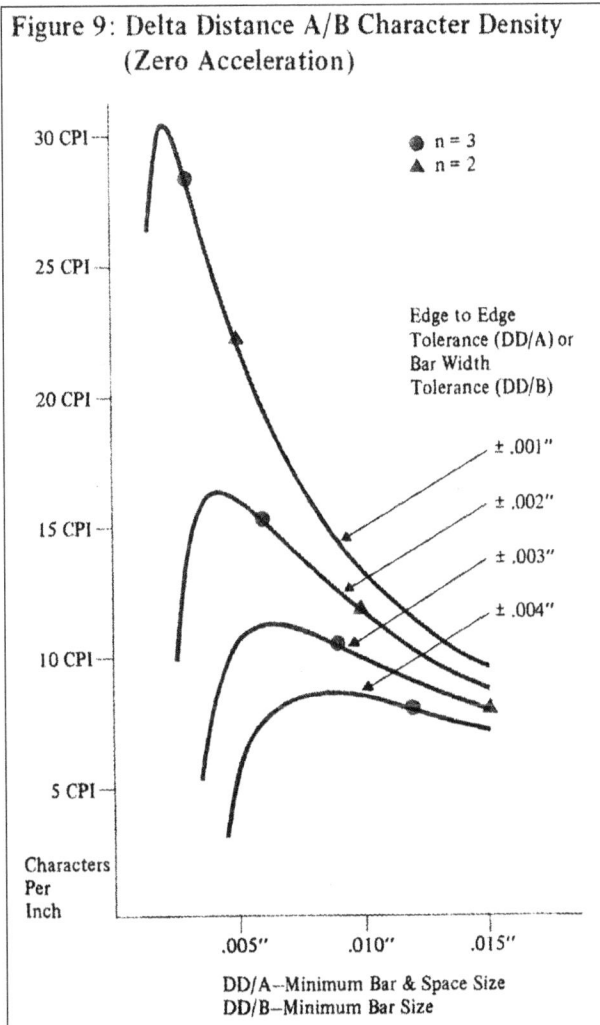

Figure 9: Delta Distance A/B Character Density (Zero Acceleration)

changes with hand-held scanner devices. The other codes do, however, meet the general criteria.

A convenient system of evaluation of relative efficiency is the character length expressed in time slot units. This comparison between the two color codes is shown in Figure 7. As shown, the Delta Distance B code is the shortest in terms of time slots. It is, however, true that the method of reading (bar-width versus edge-to-edge) is also a determining factor in achievable character density.

"Bar-width" means simply that the information is conveyed to the scanner by reading the width of a bar or space. Delta Distance B is this type of code. "Edge-to-edge" means that information is conveyed by reading from the edge of one bar to the edge of the next bar. Delta Distance A and Delta Distance C are this type of code. Generally speaking, most impact printers will print edge-to-edge dimensions more accurately than bar-width dimensions. For this reason, the Delta Distance C code can be printed at a higher density than Delta Distance B for most printers even though Delta Distance B is slightly shorter in terms of time slot units. This will be shown later.

Delta Distance Code appears to be more efficient than other two-level codes and also provides relatively simple decoding and self-clocking. A method of analysis to determine the character density potential is shown in the following sections.

Printed Two-level Code Character Density

The achievable character density of the printed code is a function of the combination of dimensional errors of the printed bars, the errors introduced by the reading mechanism and by the operator velocity change or acceleration as the reading scanner is passed across the bar-coded character.

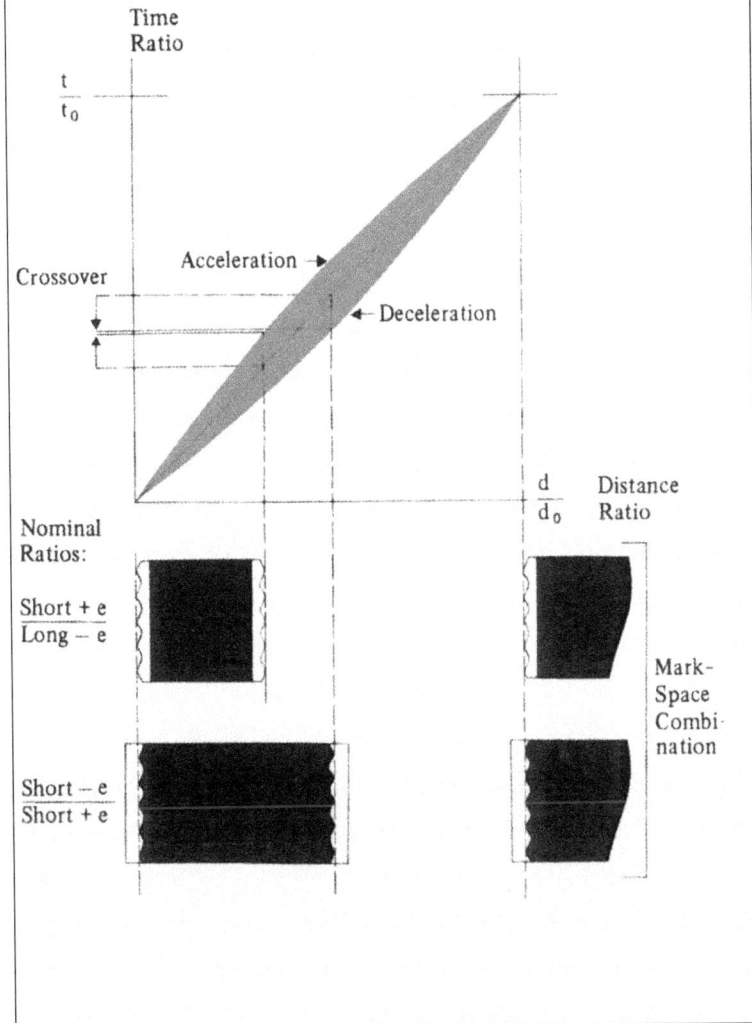

Figure 10: Acceleration & Dimensional Tolerance: Delta Distance B Code

In order to evaluate the effects of these influences, a method of analysis known as "worst case" is applied to the code. This method assumes that all tolerances reach their maximum limits simultaneously. This results in the greatest probability of character error or misread. A major disadvantage of this method of analysis is that it often results in conclusions that are more conservative than required. In actual applications, reasonable error rates are allowed and "worst cases" do not occur simultaneously, except in rare circumstances. It is, however, an accurate way of making a reasonably true comparison between codes.

The worst case situation with a DD/B type of code is shown in Figure 8. As shown, the worst case for a bar-space combination which is printed with a short-to-long ratio is for the bar to increase in width while the space decreases. It also shows that the worst case for a bar-space combination which is printed with a short-to-short ratio is for the bar to decrease in width while the space increases. When these two ratios are equal or "cross over," they are indistinguishable from each other and a character error results.

This cross-over condition can be expressed mathematically as shown at the bottom of Figure 8.

Equations 1 and 2 of Figure 8 can be solved to produce the graph shown in Figure 9, which shows the achievable character density for the Delta Distance B code under conditions of worst case dimensional tolerance.

This same analysis applies to Delta Distance A if we assume that a time slot consists of a bar and space and the printer tolerance is the edge-to-edge tolerance rather than width tolerance.

Delta Distance A and B with acceleration

The character density capability of Delta Distance A and B is influenced by the operator velocity variation of the reading scanner, in addition to dimensional tolerance. The effect of simultaneous dimensional change and acceleration is shown graphically in Figure 10.

As shown in Figure 10, the worst case for the "short-to-long" ratio is a *plus* dimensional tolerance on the bar and an acceleration by the operator across the combination. The worst case for the "short-to-short" ratio is a *minus* dimensional tolerance and a deceleration by the operator.

The mathematics for DD/A and DD/B with acceleration and dimensional variation are rather complex and are shown on Figures 11-14. Equation 6 of the analysis expresses the time taken to pass across a mark space combination of various distance ratios and under conditions of various velocity ratios. The velocity ratio as a function of operator acceleration, initial velocity and total distance is expressed by Equation 8. The distance ratio, as influenced by dimensional error, is expressed by Equations 9 and 10. The assumption is made in this analysis that acceleration or deceleration is constant across the small bar and space interval.

This set of equations can be used to express the Delta Distance B character density as a function of dimensional tolerance, minimum bar size and acceleration. This solution is shown graphically in Figure 15 for worst case acceleration.

The worst case accelerations/decelerations were established by recording operator velocities and accelerations for a number of subjects using a hand-held, pencil-like scanner device. The worst case experienced in this test at low initial velocities (10 inches/second) was about 1.7G acceleration and -.8G deceleration within a range of 5" of start (G = 386 inches/second²).

These are rather extreme cases when it is considered that this deceleration at 10 inches/second will bring the scanner to a complete stop in .16" motion.

As shown by comparing Figures 9 and 15, the character density for Delta Distance B is largely determined by the minimum bar size and the dimensional tolerance. The operator acceleration is a relatively minor influence at reasonable character densities.

The same analysis applies to Delta Distance A if it is assumed that a time slot consists of a bar and space and the printer tolerance is the edge-to-edge tolerance rather than width tolerance.

Delta Distance C Analysis

Delta Distance C is similar to Delta Distance B, as indicated earlier, except that it has the property of being read from adjacent bar edges rather than across the width of a bar.

The worst case tolerances for this type of code are shown in Figure 16.

The major difference between Delta Distance C and Delta Distance B is that the bars are read edge-to-edge to establish four time bases (T_1 through T_4 in Figure 16), which are then divided by a constant reference ($T_1 + T_3$) to divide the code into a series of long, medium and short time bases. The pattern of longs, mediums, and shorts defines the character and, via a logical algorithm, can be converted to the binary value of the character. The worst case tolerance condition for this character is different from Delta Distance B because of the fact that the length of the time bases are derived with respect to the constant width reference. The worst case occurs when

Figure 11: Acceleration Equation: Mark-Space Pair

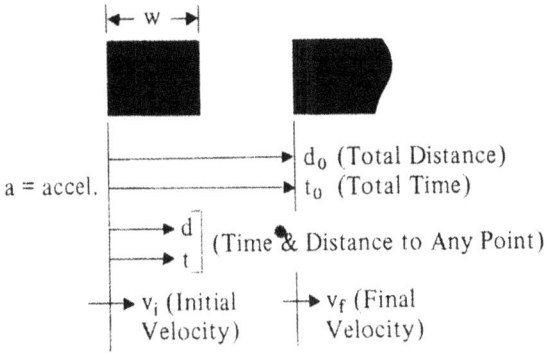

a = accel.

d_0 (Total Distance)

t_0 (Total Time)

d / t (Time & Distance to Any Point)

v_i (Initial Velocity) v_f (Final Velocity)

Assume a = Constant

$$d = \frac{1}{2}at^2 + v_i t$$

Normalize:

$$\frac{d}{d_0} = \frac{\frac{at_0}{2d_0}}{t_0}\left(\frac{t}{t_0}\right)^2 + \frac{\frac{v_i}{d_0}}{t_0}\left(\frac{t}{t_0}\right) \qquad 1$$

Sub in Equation 1:

$$\frac{d_0}{t_0} = \frac{v_f + v_i}{2} \quad \& \quad at_0 = v_f - v_i \qquad 2, 3$$

$$\therefore \frac{d}{d_0} = \frac{v_f - v_i}{\frac{2(v_f + v_i)}{2}}\left(\frac{t}{t_0}\right)^2 + \frac{2v_i}{v_f + v_i}\left(\frac{t}{t_0}\right)$$

Which simplifies to:

$$\frac{d}{d_0} = \frac{1 - v_r}{1 + v_r}\left(\frac{t}{t_0}\right)^2 + \frac{2v_r}{1 + v_r}\left(\frac{t}{t_0}\right) \qquad 4$$

$$\text{in which } v_r = \frac{v_i}{v_f} \qquad 5$$

the medium time base becomes indistinguishable from or "crosses over" with the long time base. This condition is expressed mathematically at the bottom of Figure 16.

The character density for Delta Distance C using this analysis is shown in Figure 17. This may be compared directly with the Delta Distance B density shown in Figure 9 for the zero acceleration case.

Figure 12

Equation 4 of Figure 11 can be solved for $\dfrac{t}{t_0}$ using the Quadratic Formula.

Let $\dfrac{t}{t_0} = t_r$ $\dfrac{d}{d_0} = d_r$

$$t_r = \frac{-\dfrac{2v_r}{1 + v_r} \pm \sqrt{\dfrac{4v_r^2}{(1 + v_r)^2} - 4\left(\dfrac{1 - v_r}{1 + v_r}\right) d_r}}{2 \left(\dfrac{1 - v_r}{1 + v_r}\right)}$$

Which simplifies to:

$$t_r = \frac{-v_r}{1 - v_r} \pm \sqrt{\frac{v_r^2}{1 - v_r^2} - \left(\frac{1 + v_r}{1 - v_r}\right) d_r} \qquad 6$$

Note: "+" Applies to $v_r < 1$ (Accel)
         "−" Applies to $v_r > 1$ (Decel)

Figure 13

Express v_r as $f(a, d_0, v_i)$
from Equation 2 in Figure 11:

$$t_0 = \frac{2d_0}{v_f + v_i} \qquad 7$$

Sub 7 in Equation 3:

$$a\left(\frac{2d_0}{v_f + v_i}\right) = v_f - v_i$$

$$2ad_0 = v_f^2 - v_i^2$$

$$\frac{2ad_0}{v_i^2} = \frac{v_f^2}{v_i^2} - 1$$

$$\frac{v_f^2}{v_i^2} = \frac{2ad_0 + v_i^2}{v_i^2}$$

$$v_r = \sqrt{\frac{v_i^2}{v_i^2 + 2ad_0}} \qquad 8$$

Figure 14

Express Distance Ratio as function of
Dimensional Variation:

Let: w = Smallest Mark Width (Time Slot)

e = Dimensional Error across width of Mark

n = Long to Short Ratio

Then:

d_r for Short to Long Ratio (Worst Case)

$$d_r = \frac{w + e}{(n + 1)\, w} \qquad 9$$

d_r for Short to Short Ratio (Worst Case)

$$d_r = \frac{w - e}{2w} \qquad 10$$

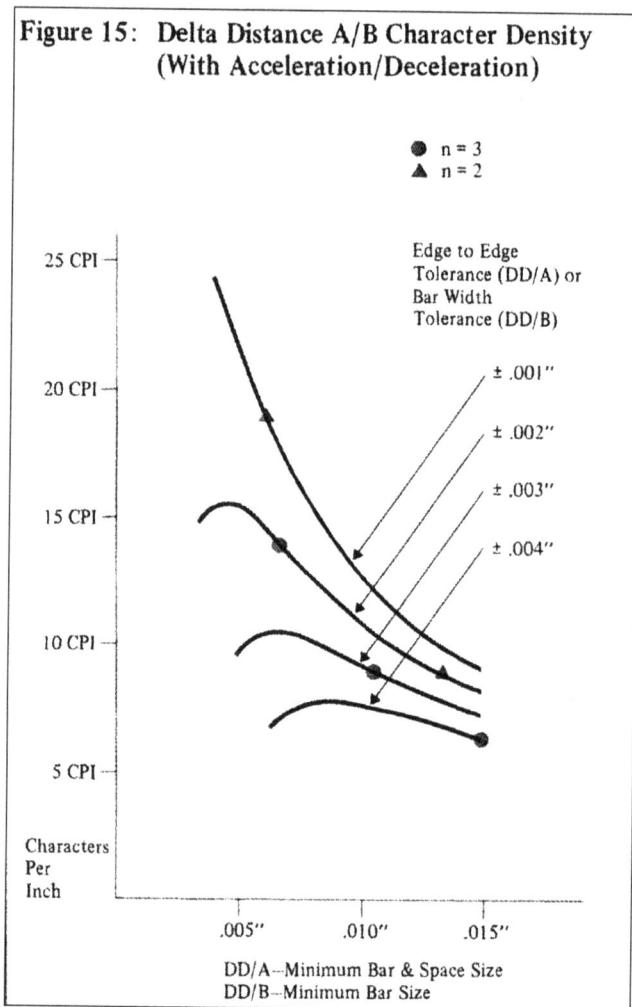

Figure 15: Delta Distance A/B Character Density (With Acceleration/Deceleration)

Figure 16: Tolerance--Delta Distance C Code

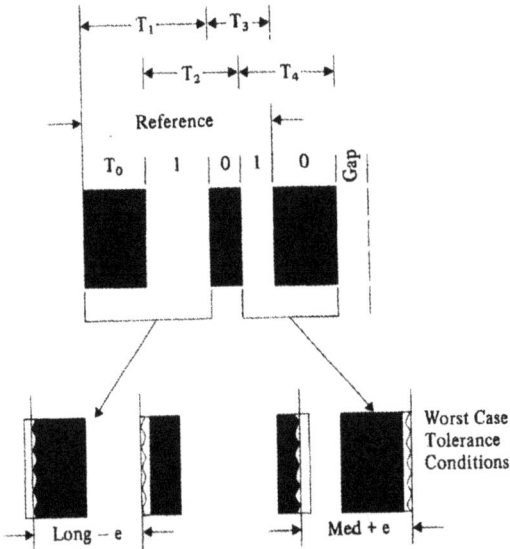

DD/C is decoded by dividing the Time Bases $(T_1 - T_4)$ by the Constant Width Reference. The Worst Case occurs when the "Long" Time Bases "Crossover" with the "Medium" Time Bases. Mathematically this is:

$$\frac{M}{Constant} = \frac{L}{Constant}$$

$$\frac{.(n + 1)\,w + e}{(2 + 2n)\,w - e} = \frac{(2n)\,w - e}{(2 + 2n)\,w + e}$$

in which: n = Long to Short Bar Ratio
          e = Dimensional Error
          w = Minimum Bar Width

Which simplifies to: $e = \left[\dfrac{2n^2 - 2}{7n + 5}\right] w$

The Character Length is: $L = (3 + 3n)\,w + 1$
in which L = Length
          I = Character-to-character tolerance

APPENDIX

Delta Distance C with Acceleration

The character density of Delta Distance C is affected by operator acceleration in a similar way to Delta Distance B. Note, however, in Figure 18 that the acceleration or deceleration occurs simultaneously over both the time base which is being read and the reference time base within the character. The equations for the crossover condition of Delta Distance C with acceleration are developed in a similar fashion to that shown in Figures 11 through 14 and the solution for various tolerances and bar sizes is shown in Figure 19.

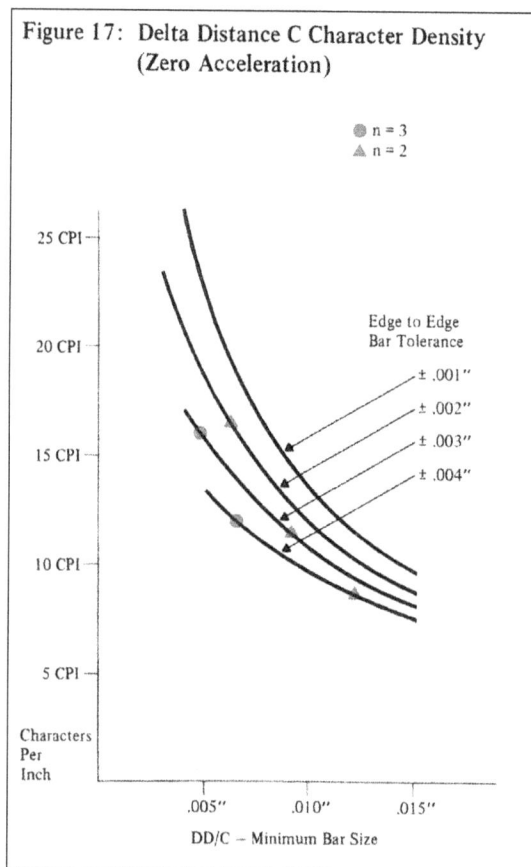

Figure 17: Delta Distance C Character Density (Zero Acceleration)

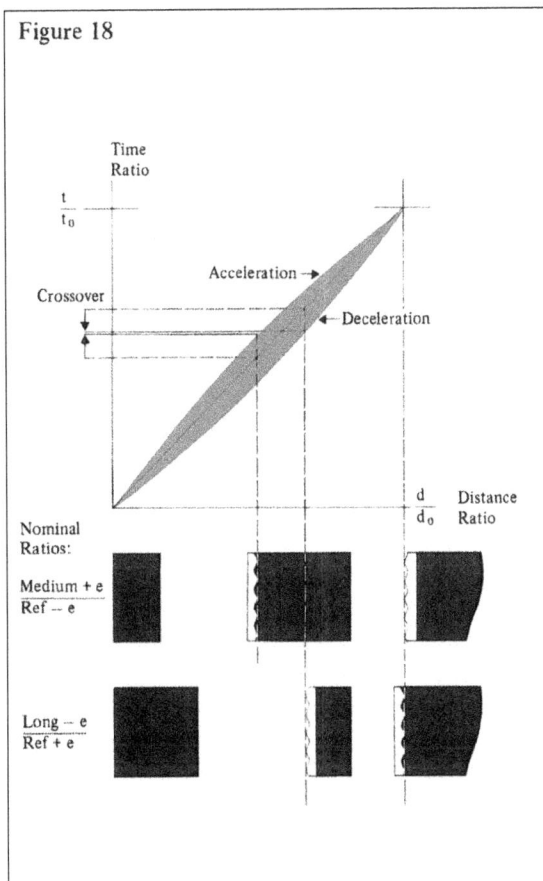

Figure 18

Operational Density of Printed Delta Distance Code

The character density of printed Delta Distance A, B, and C which can be achieved in practice is determined by several variables:

1. The accuracy of the printer

2. The accuracy of the reader

3. The minimum bar size

4. The operator velocity change or acceleration

An example of the character densities which can be achieved in practice with Delta Distance is shown below based on the tolerances associated with MICR (bank font)-type printers.

Printers will vary over a considerable range of accuracies. However, if we assume tolerances based on the printers which presently print MICR-type characters, there is a considerable body of information available on achievable character line tolerances. Based on MICR tolerances, the assumed tolerances for DD/B and DD/C are:

	Bar-width	Edge-to-edge	Based on
DD/B	±.0030"		E13B(MICR)
DD/A & DD/C		±.0016"	CMC7(MICR)

In addition to printer tolerances, we must also include a reader tolerance. This is a complicated function of aperture size, aperture orientation, edge conditions, etc. Experimental evidence indicates, however, that optical or magnetic reader tolerances are considerably smaller than the printer tolerances and, for hand-held optical-type readers, tend to fall in the following range:

	Reader error
Bar-width	±.00048"
Edge-to-edge	±.00024"

The bar-width error tends to be higher than the edge-to-edge error, because the reading threshold must be set differently for the leading and trailing edge of the bar. These thresholds tend to be slightly different, which introduces the additional error. The other limit on character density is minimum bar size. If we again assume that MICR establishes these minimum bar sizes, then this limit is:

	Minimum bar size	Based on
Edge-to-edge	.0057"	CMC7 (MICR)
Bar-width	.0110"	E13B (MICR)

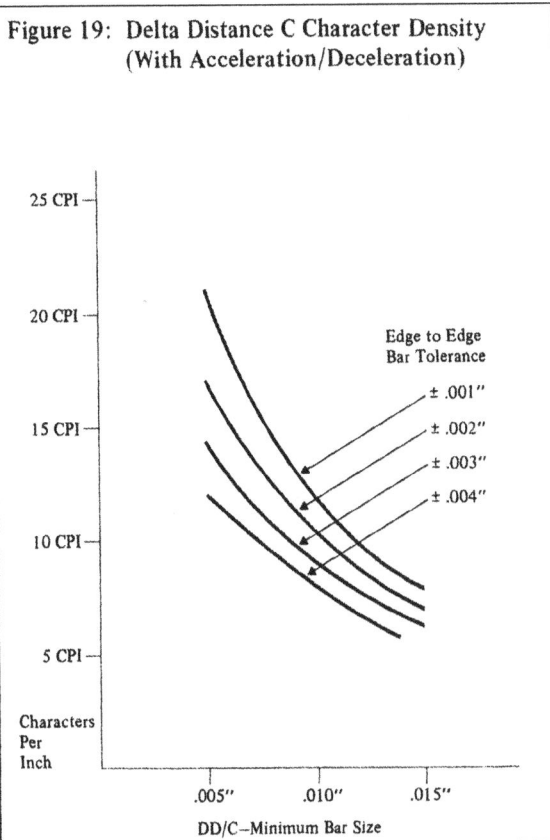

Figure 19: Delta Distance C Character Density
(With Acceleration/Deceleration)

Entering these tolerances and limits on the graphs shown in Figures 9, 15, 17 and 19 results in the following limits for character density:

| | Characters/inch (numeric) | | Total | Minimum |
	With accel	W/O accel	tolerance	bar/space
DD/A	10.1	10.7	±.00184	.0114 (Bar & space)
DD/B	8.1	9.0	±.00348	.0110 (Bar only)
DD/C	16.1	17.6	±.00184	.0057 (Bar only)
(Acceleration = +1.8G/−.7G; Char-Char Tolerance = ±.005″)				

It should be kept in mind that the character density fig-ures shown for printed bar codes are based on a worst case analysis and on present working tolerances for MICR print-ers. An analysis of most codes, including the three-level bar code, by this method will yield character densities consid-erably below those achieved in practice. For this reason, the densities shown above should be compared only with densi-ties resulting from a worst case analysis of other codes.

Operational Density of Magnetic Delta Distance Code

Most codes can be recorded on magnetic media more accu-rately than they can be printed in bar form.

As previously shown, the achievable character density of a printed bar code is largely a function of the printer's accuracy.

In recorded code, the reader accuracy and magnetic me-dia characteristics generally govern the achievable density. In the case of hand-held magnetic readers, this includes such effects as reader skew and operator velocity change.

For this reason, it is more difficult to make general state-ments about achievable densities with magnetic encoding. Generally speaking, however, the achievable character den-sities of magnetic encoding when read with fixed velocity transports and well-aligned reading heads tend to be greater than that of printed bar codes. With hand-held magnetic readers which are subject to reading under conditions of reader skew and operator velocity change, the magnetic den-sity still tends to be somewhat greater than printed bar code density, but somewhat less than the density of magnetic me-dia read in a fixed velocity transport.

Decoding and detection

The detection and decoding of Delta Distance codes will generally be accomplished by converting the signal from the reading element (the magnetic head or the optical photo element) into a digital "square wave" which is an electrical representation of the bar. Once converted, this digital signal may be counted and stored in digital registers and compared with "counts" from the previous intervals within the code to derive the bit value of the code.

With modern integrated circuits and programmable devices such as might be found in a point-of-sale terminal, the difference in decoding logic between Delta Distance A, B or C, or any other self-clocking code, is insignificant.

The decoding logic for a given code is the same whether the information was read optically or magnetically, once the signal has been converted to a digital square wave by the detector. This means that magnetic and optical reading devices might be "mixed" in a given system and still use the same decoding logic.

The essential element is that a common *code* be chosen for all media—magnetic or optical.

Summary

The code appears to be the key to the problem of future merchandise and credit identification, since it largely determines the characteristics of the other elements of the system. An optimum code would seem to be one which is both recordable with conventional magnetics and printable with conventional types of single-color impact printers.

The Delta Distance Code meets this requirement and provides both character density efficiency and relatively simple decoding.

Delta Distance C is the more efficient version for today's type of impact, character-by-character printers. Delta Distance B is slightly more efficient for magnetic stripe recording, since it contains one less time slot per character than Delta Distance C. However, since code length is generally not a limiting factor in magnetic recording, Delta Distance C is the best choice of a code which is equally applicable to printed bar code and magnetically recorded code.

A sample of Delta Distance C as it appears in the printed version is shown in Figure 20.

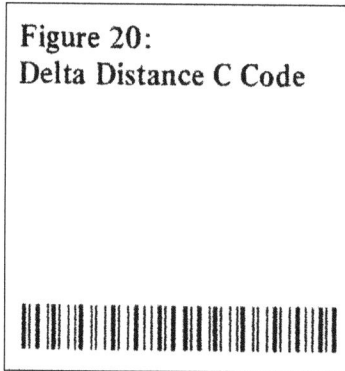

Figure 20:
Delta Distance C Code

ABOUT THE AUTHOR

Paul V. McEnroe is an award-winning engineer who developed multiple state-of-the-art technologies during his long career, including more than two decades in leadership roles at IBM. He grew up in Ohio, was valedictorian at the University of Dayton, and earned advanced degrees in engineering from Purdue University and Stanford University. He also completed executive programs at UCLA and Northwestern's Kellogg School of International Management. He was named the University of Dayton's Most Distinguished Alumnus in 1999, and later received several coveted awards from Purdue, and an honorary Doctor of Science from California State University and California Polytechnic State University, San Luis Obispo.

McEnroe is best known for his primary role in creating the Universal Product Code (UPC), the barcode used on every product in supermarkets and the retail industry, and the scanners that read them. In 1974 he received the IBM SCD President's Award for the development of the Supermarket System including the Barcode ". . . from inception through shipment." He was co-inventor of the handheld "pistol-grip" scanner that reads barcodes from a distance, developed the magnetic code for stock keeping unit (SKU) marking, and later managed the development of the Token Ring Local Area Network (LAN). In 1984 McEnroe became president of Trilogy Systems Corporation and developed other multi-chip technologies.

A lifelong equestrian enthusiast, McEnroe, with his wife Tina, own and operate a horse breeding and boarding facility and working cattle ranch in California's Santa Ynez Valley. He still competes in Rodear (equestrian skill) competitions, has served as president and state director of the Santa Barbara County Cattlemen's Association, and was named Livestock Person of the Year for Santa Barbara County. McEnroe was honored as Vaquero of the Year in 2022.

In 2011 Paul and Tina founded the McEnroe Reading & Language Arts Clinic at the University of California, Santa Barbara. Paul has served as advisor or director/trustee of several other nonprofits in the Central Coast community including the Land Trust for Santa Barbara County; St. John's Seminary College, Camarillo; and, for thirty-seven years, the Cal Poly President's Council.

www.ingramcontent.com/pod-product-compliance
Lightning Source LLC
Chambersburg PA
CBHW040754220326
41597CB00029BA/4811